The Core Model Iterability Problem

Since their inception, the Perspectives in Logic and Lecture Notes in Logic series have published seminal works by leading logicians. Many of the original books in the series have been unavailable for years, but they are now in print once again.

Large cardinal hypotheses play a central role in modern set theory. One important way to understand such hypotheses is to construct concrete, minimal universes, or "core models", satisfying them. Since Gödel's pioneering work on the universe of constructible sets, several larger core models satisfying stronger hypotheses have been constructed, and these have proved quite useful. In this volume, the 8th publication in the Lecture Notes in Logic series, Steel extends this theory so that it can produce core models having Woodin cardinals, a large cardinal hypothesis that is the focus of much current research. The book is intended for advanced graduate students and researchers in set theory.

JOHN R. STEEL works in the Department of Mathematics at the University of California, Berkeley.

LECTURE NOTES IN LOGIC

A Publication of The Association for Symbolic Logic

This series serves researchers, teachers, and students in the field of symbolic logic, broadly interpreted. The aim of the series is to bring publications to the logic community with the least possible delay and to provide rapid dissemination of the latest research. Scientific quality is the overriding criterion by which submissions are evaluated.

More information, including a list of the books in the series, can be found at http://www.aslonline.org/books-lnl.html

LECTURE NOTES IN LOGIC 8

The Core Model Iterability Problem

JOHN R. STEEL
University of California, Berkeley

ASSOCIATION FOR SYMBOLIC LOGIC

CAMBRIDGE
UNIVERSITY PRESS

University Printing House, Cambridge CB2 8BS, United Kingdom

One Liberty Plaza, 20th Floor, New York, NY 10006, USA

477 Williamstown Road, Port Melbourne, VIC 3207, Australia

4843/24, 2nd Floor, Ansari Road, Daryaganj, Delhi – 110002, India

79 Anson Road, #06–04/06, Singapore 079906

Cambridge University Press is part of the University of Cambridge.

It furthers the University's mission by disseminating knowledge in the pursuit of education, learning, and research at the highest international levels of excellence.

www.cambridge.org
Information on this title: www.cambridge.org/9781107167964
10.1017/9781316716892

Association for Symbolic Logic
Richard A. Shore, Publisher
Department of Mathematics, Cornell University, Ithaca, NY 14853
http://www.aslonline.org

A catalogue record for this publication is available from the British Library.

ISBN 978-1-107-16796-4 Hardback

Table of Contents

The Core Model Iterability Problem

J. R. Steel

These notes develop a method for constructing *core models*, that is, canonical inner models of the form $L[\boldsymbol{E}]$, where $\boldsymbol{E}$ is a coherent sequence of extenders. They extend the earlier work in this area of Dodd and Jensen ([DJ1], [DJ2], [DJ3]) and Mitchell ([M1], [M?]). The Dodd-Jensen theory produces models having measurable cardinals, and Mitchell's extension of it produces models having measurable cardinals κ of order κ^{++}. Here we shall extend this theory so that it can produce core models having Woodin cardinals.

The extent of our debt to Dodd, Jensen, and Mitchell will become apparent; nevertheless, we shall not assume that the reader is familiar with their work. We shall, however, assume that he is familiar with the fine structure theory for core models having Woodin cardinals which is developed in [FSIT].

Our work here goes beyond [FSIT] in that it involves a construction of $L[\boldsymbol{E}]$ models which makes no use of extenders over V. Many of the applications of core model theory require such a construction. The authors of [FSIT] use extenders over V in order to show that the inner model $L[\boldsymbol{E}]$ they construct is sufficiently iterable: roughly, they demand that the extenders put onto $\boldsymbol{E}$ be the restrictions of background extenders over V, then use this fact to embed iteration trees on $L[\boldsymbol{E}]$ into iteration trees on V, and then quote the results of [IT] concerning iteration trees on V. Here we shall describe a weakened background condition on the extenders put onto $\boldsymbol{E}$ which does not require full extenders over V, and yet suffices to carry out something like the old proof of the iterability of $L[\boldsymbol{E}]$. The result is a solution to what is called the "core model iterability problem" in [FSIT].

The notes are organized as follows. As in Mitchell's work on the core model for sequences of measures ([M1], [M?]), we construct the model K in which we are ultimately most interested in two steps. In §1 we construct a model, which we call K^c, whose extenders have "background certificates". These background certificates guarantee the iterability of K^c, its levels, and various associated structures. (In Mitchell's work, the background condition is countable completeness, but here we seem to need more.) In order to show that K^c and the K we derive from it are large enough to be useful, we seem to need something like the existence of a measurable cardinal. We fix a normal measure μ_0 on a measurable cardinal Ω throughout this paper; we shall have $\mathrm{OR} \cap K^c = \mathrm{OR} \cap K = \Omega$. We use the measurability of Ω to show in §1 that either $K^c \models$ there is a Woodin cardinal, or $(\alpha^+)^{K^c} = \alpha^+$ for μ_0- a.e. $\alpha < \Omega$. Since in applications we are seeking an inner model with a Woodin cardinal, we assume through most of the paper that there is no such model, and thus we have $(\alpha^+)^{K^c} = \alpha^+$ for μ_0- a.e. $\alpha < \Omega$. As in Mitchell's work, this "weak covering property" of K^c is crucial.

We also use the measurability of Ω in a different way in §4. Further, we use the tree property of Ω to show that iteration trees of length Ω are well behaved. The fact that we do not develop the basic theory of K within ZFC may be a defect in our work.

In §2 we sketch the main new ideas in the proof that K^c is iterable. In §9 we give a full proof of a general iterability theorem which covers iteration trees and psuedo-iteration trees on K^c, its levels, and the associated bicephali and psuedo-premice.

As in Mitchell's work, the "true core model" K is a Skolem hull of K^c. In §3 and §4 we develop some concepts, derived from Mitchell's work, which are useful in the construction of this hull. In §5 we do the construction: given a stationary $S \subseteq \Omega$ with certain properties, we construct a model $K(S) \preceq K^c$. We show $(\alpha^+)^{K(S)} = \alpha^+$ for μ_0- a.e. α. We also show that $K(S)$ is invariant under small forcing; that is, $K(S) = K(S)^{V[G]}$ whenever G is V generic over some $\mathbb{P} \in V_\Omega$. Finally, we show that $K(S)$ is independent of S, and define K to be the common value of $K(S)$ for all S. We have then that $(\alpha^+)^K = \alpha^+$ for μ_0 - a.e. α, and that K is invariant under set forcing.

In §6 we give an optimally simple inductive definition of K: it turns out that $K \cap HC$ is $\Sigma_1(L_{\omega_1}(\mathbb{R}))$. (Woodin has shown that no simpler definition is possible in general. Mitchell showed in [M?] that if no initial segment of K satisfies $(\exists\kappa)(o(\kappa) = \kappa^{++})$, then $K \cap HC$ is Σ^1_5 in the codes.) In §7 we use the machinery we have developed to obtain the consistency strength lower bound of one Woodin cardinal for various propositions. In §8 we return to the pure theory, and obtain some information concerning embeddings of K. We show, for example, that if there is no inner model with a Woodin cardinal, then there is no nontrivial elementary $j : K \to K$. In contrast to the situation for "smaller K's", however, we show that there may be nontrivial elementary $j : K \to M$ which are not iteration maps.

Among the applications of the theory developed in these notes are the following theorems.

Theorem 0.1. *Let Ω be measurable, and suppose there is a presaturated ideal on ω_1; then there is a transitive set $M \subseteq V_\Omega$ such that*

$$M \models ZFC + \text{"There is a Woodin cardinal"}.$$

Corollary 0.2. *If Martin's Maximum holds, then there is a transitive set M such that*

$$M \models ZFC + \text{"There is a Woodin cardinal"}.$$

(It is known from [FMS] that Martin's Maximum implies that there is an inner model with a measurable cardinal and a saturated ideal on ω_1; by applying 0.1 inside this model we get 0.2. H. Woodin pointed this out to the author.)

Further work of Mitchell, Schimmerling, and the author on the weak covering property for K (cf. [WCP]) together with his work on Jensen's $\square$ principle in K (cf. [Sch]), led Schimmerling to the following improvement of 0.2.

Theorem 0.3. (Schimmerling, cf. [Sch]) *If PFA holds, then there is a transitive set M such that*

$$M \models ZFC + \text{"There is a Woodin cardinal"}.$$

(Theorem 0.3 also relies on an improvement, due to Magidor, of Todorčević's result that PFA implies $\forall\kappa$ ($\Box_\kappa$ fails). (See [To].)

In a different vein, we have the following, more immediate applications of the theory presented here.

Theorem 0.4. *Suppose that every set of reals which is definable over $L_{\omega_1}(\mathbb{R})$ is weakly homogeneous; then there is a transitive set M such that*

$$M \models ZFC + \text{"There is a Woodin cardinal"}.$$

(H. Woodin supplied a crucial step in the proof of 0.4.) Since Woodin (unpublished) has shown that the existence of a strongly compact cardinal implies the hypothesis of 0.4, we have

Corollary 0.5. *Suppose there is a strongly compact cardinal; then there is a transitive set M such that*

$$M \models ZFC + \text{"There is a Woodin cardinal"}.$$

(We shall give a different proof of 0.5, one which avoids Woodin's unpublished work, in §8.)

We can also improve the lower bound of [IT] on the strength of the failure of UBH.

Theorem 0.6. *Let Ω be measurable, and suppose there is an iteration tree $\mathcal{T}$ on V_Ω such that $\mathcal{T} \in V_\Omega$ and $\mathcal{T}$ has distinct cofinal wellfounded branches; then there is a transitive set $M \subseteq V_\Omega$ such that*

$$M \models ZFC + \text{"There are two Woodin cardinals"}.$$

We can use the methods presented here to re-prove Woodin's result that $\forall x \in^\omega \omega(x^\sharp$ exists$) + \Delta^1_2$ determinacy implies that there is an inner model with a Woodin cardinal. Finally, using an idea of G. Hjorth, the author has recently (at least partially) generalized Jensen's Σ^1_3 correctness theorem to the core model constructed here. This yields a positive answer to a conjecture of A. S. Kechris.

Theorem 0.7. *Assume $\forall x \in {}^\omega\omega$ ($x^\sharp$ exists and $\Sigma^1_3(x)$ has the separation property); then there is a transitive set M such that*

$$M \models ZFC + \text{"There is a Woodin cardinal"}.$$

Each of the hypotheses of the theorems above is known to be consistent under some large cardinal hypothesis or other. We shall not attempt a scholarly discussion of the history of or context for these theorems, as our focus here is the basic theory which produces them. We shall prove 0.1, 0.4, 0.6, and 0.7 in §7.

Historical note. We did most of the work described here in the Spring of 1990, and informally circulated it in a set of handwritten notes. Our main advance, which was isolating K^c and proving the results of §1 and §2 concerning it, was inspired in part by some ideas of Mitchell. The work in §8 was done somewhat later, and the Σ^1_3 correctness theorem of §7D was not proved until the Spring of 1993.

§1. The construction of K^c

Let us fix a measurable cardinal, which we call Ω, for the remainder of this paper. We shall sometimes think of Ω as the class of all ordinals; we could have worked in 3rd order set theory + "OR is measurable", but opted for a little more room. Fix also a normal measure μ_0 on Ω.

We now define by induction on $\xi < \Omega$ premice $\mathcal{N}_\xi$. Having defined $\mathcal{N}_\xi$, we set

$$\mathcal{M}_\xi = \mathfrak{C}_\omega(\mathcal{N}_\xi),$$

the ωth core of $\mathcal{N}_\xi$. The $\mathcal{M}_\xi$'s will converge to the levels of K^c, the "background-certified" core model for one Woodin cardinal. That is, we shall define K^c by setting

$$\mathcal{J}_\beta^{K^c} = \text{eventual value of } \mathcal{J}_\beta^{\mathcal{M}_\xi}, \quad \text{as } \xi \to \Omega,$$

for all $\beta < \Omega$. The construction of the $\mathcal{N}_\xi$'s follows closely the construction in §11 of [FSIT].

In this section, we shall simply assume that the $\mathcal{N}_\xi$'s are all "reliable", that is, that $\mathfrak{C}_k(\mathcal{N}_\xi)$ exists and is k-iterable for all $k \leq \omega$. By Theorem 8.1 of [FSIT], this amounts to assuming that if $\mathfrak{C}_k(\mathcal{N}_\xi)$ exists, then it is k-iterable (for all $k \leq \omega$). We shall sketch a proof of this iterability assumption in §2, and give a full proof in §9. We shall also assume here that certain bicephali and psuedo-premice associated to $\langle \mathcal{N}_\xi \mid \xi < \Omega \rangle$ are sufficiently iterable. We prove this in §9.

Iterability comes from the existence of background extenders.

Definition 1.1. *Let $\mathcal{M}$ be an active premouse, F the extender coded by $\dot{F}^{\mathcal{M}}$ (i.e. its last extender), $\kappa = crit(F)$, and $\nu = \dot{\nu}^{\mathcal{M}} = \sup$ of the generators of F. Let $\mathcal{A} \subseteq \bigcup_{n<\omega} P([\kappa]^n)^{\mathcal{M}}$. Then an $\mathcal{A}$-certificate for $\mathcal{M}$ is a pair (N, G) such that*

(a) *N is a transitive, power admissible set, $V_\kappa \cup \mathcal{A} \subseteq N$, ${}^\omega N \subseteq N$, and G is an extender over N,*

(b) *$F \cap ([\nu]^{<\omega} \times \mathcal{A}) = G \cap ([\nu]^{<\omega} \times \mathcal{A})$,*

(c) *$V_{\nu+1} \subseteq Ult(N, G)$, and ${}^\omega Ult(N, G) \subseteq Ult(N, G)$,*

(d) *$\forall \gamma(\omega\gamma < OR^{\mathcal{M}} \Rightarrow \mathcal{J}_\gamma^{\mathcal{M}} = \mathcal{J}_\gamma^{i(\mathcal{J}_\kappa^{\mathcal{M}})})$, where $i : N \to Ult(N, G)$ is the canonical embedding.*

Definition 1.2. *Let $\mathcal{M}$ be an active premouse, and κ the critical point of its last extender. We say $\mathcal{M}$ is countably certified iff for every countable $\mathcal{A} \subseteq \bigcup_{n<\omega} P([\kappa]^n)^{\mathcal{M}}$, there is an $\mathcal{A}$-certificate for $\mathcal{M}$.*

In the situation described in 1.2 we shall typically have $|N| = \kappa$, so that $(\mathrm{OR} \cap N) < lhG$. We are therefore not thinking of (N, G) as a structure to be iterated; N simply provides a reasonably large collection of sets to be measured by G. The conditions $V_\kappa \subseteq N$ and $V_{\nu+1} \subseteq \mathrm{Ult}(N, G)$ are crucial; the closure of N and $\mathrm{Ult}(N, G)$ under ω-sequences could probably be dropped.

It might seem that certificates (N, G) as in 1.1 are too much to ask for, and in particular condition 1.1 (c) might seem too strong. But notice that we get such pairs by taking Skolem hulls: if $\pi : N \cong H \prec V_\eta$ inverts the collapse of H, where $V_\kappa \subseteq H$ but $\kappa \notin H$, then letting G be the length $\pi(\kappa)$ extender derived from π, G is an extender over N, $V_\kappa \subseteq N$, and $V_{\pi(\kappa)} \subseteq \mathrm{Ult}(N, G)$. We shall also see that the embedding associated to the measure μ_0 on Ω gives rise to certificates.

Definitions 1.1 and 1.2 were inspired in part by earlier attempts by Mitchell to formulate a background condition along these lines.

We proceed to the inductive definition of $\mathcal{N}_\xi$. As we define the $\mathcal{N}_\xi$'s we verify:

A_ξ : $\forall\alpha < \xi\ \forall\kappa$ (if $\rho_\omega(\mathcal{M}_\gamma) \geq \kappa$ for all γ s.t. $\alpha < \gamma \leq \xi$, then letting $\eta = (\kappa^+)^{\mathcal{M}_\alpha}$, $\mathcal{J}_\eta^{\mathcal{M}_\alpha} = \mathcal{J}_\eta^{\mathcal{M}_\xi}$).

(Here we let $\omega\eta = \mathrm{OR} \cap \mathcal{M}_\alpha$ in the case $\mathcal{M}_\alpha \models \kappa^+$ doesn't exist.)

We begin by setting $\mathcal{N}_0 = (V_\omega, \in)$. (So $\mathcal{M}_0 = \mathcal{N}_0$.) Now suppose $\mathcal{N}_\xi$, and hence $\mathcal{M}_\xi$, is given.

Case 1. $\mathcal{M}_\xi = (J_\gamma^{\boldsymbol{E}}, \in, \boldsymbol{E} \restriction \gamma)$ is a passive premouse, and there is an extender F over $\mathcal{M}_\xi$ such that

(1) $(J_\gamma^{\boldsymbol{E}}, \in, \boldsymbol{E} \restriction \gamma, \tilde{F})$ is a 1-small, countably certified premouse, and, letting $\kappa = \mathrm{crit}(F)$, we have

(2) κ is inaccessible,

(3) $(\kappa^+)^{\mathcal{M}_\xi} = \kappa^+ \Rightarrow$ for stationary many $\beta < \kappa$, β is inaccessible and $(\beta^+)^{\mathcal{M}_\xi} = \beta^+$.

In this case, we choose an F as above with $\nu(F)$, the sup of the generators of F, as small as possible, and we set

$$\mathcal{N}_{\xi+1} = (J_\gamma^{\boldsymbol{E}}, \in, \boldsymbol{E} \restriction \gamma, \tilde{F}).$$

As we mentioned above, the results of §9 and §8 of [FSIT] imply that $\mathcal{N}_{\xi+1}$ is reliable. Thus $\mathcal{M}_{\xi+1} = \mathfrak{C}_\omega(\mathcal{N}_{\xi+1})$ exists. We get $A_{\xi+1}$ from (the proof of) Theorem 8.1 of [FSIT].

Case 2. Otherwise.

In this case, let $\omega\gamma = \mathrm{OR} \cap \mathcal{M}_\xi$, and set

$$\mathcal{N}_{\xi+1} = (J_{\gamma+1}^{\dot{E}^{\mathcal{M}_\xi} \frown \dot{F}^{\mathcal{M}_\xi}}, \in, \dot{E}^{\mathcal{M}_\xi} \frown \dot{F}^{\mathcal{M}_\xi}).$$

Again, $\mathcal{N}_{\xi+1}$ is reliable by §9, and 8.1 of [FSIT] yields $A_{\xi+1}$.

So in Case 1 we add a countably certified extender to our extender sequence, while in Case 2 we take one step in the constructible closure of the sequence we have. In both cases we then form the ωth core of the structure we have.

Now suppose we have defined $\mathcal{N}_\xi$ for $\xi < \lambda$, where $\lambda < \Omega$ is a limit ordinal. Set

$$\eta = \liminf_{\xi \to \lambda} (\rho_\omega(\mathcal{M}_\xi)^{+^{\mathcal{M}_\xi}})$$

(where $\rho_\omega(\mathcal{M}_\xi)^{+^{\mathcal{M}_\xi}} = \mathrm{OR} \cap \mathcal{M}_\xi$ is possible. We set $\mathcal{N}_\lambda$ = unique passive premouse $\mathcal{P}$ s.t.

(a) $\forall\beta < \eta(\mathcal{J}_\beta^{\mathcal{P}}$ = eventual value of $\mathcal{J}_\beta^{\mathcal{M}_\xi}$ as $\xi \to \lambda)$ and (b) $\mathcal{J}_\eta^{\mathcal{P}} = \mathcal{P}$. Such a premouse exists as A_ξ holds for all $\xi < \lambda$. It is easy to verify A_λ.

This completes the inductive definition of the $\mathcal{N}_\xi$, for $\xi < \Omega$.

Theorem 8.1 of [FSIT] actually gives the following strengthening of our induction hypothesis A_η. (Cf. 11.2 of [FSIT].)

Lemma 1.3. *Suppose $\kappa \le \rho_\omega(\mathcal{M}_\xi)$ for all $\xi \ge \alpha_0$, and let $\xi \ge \alpha_0$ be such that $\kappa = \rho_\omega(\mathcal{M}_\xi)$. Then $\mathcal{M}_\xi$ is an initial segment of $\mathcal{M}_\eta$, for all $\eta \ge \xi$; moreover $\mathcal{M}_{\xi+1}$ satisfies "every set has cardinality $\le \kappa$".*

Lemma 1.3 implies $\liminf_{\xi\to\Omega} \rho_\omega(\mathcal{M}_\xi) = \Omega$, and therefore we can define a premouse K^c of ordinal height Ω by

$$\mathcal{J}_\beta^{K^c} = \text{eventual value of} \quad \mathcal{J}_\beta^{\mathcal{M}_\xi}, \quad \text{as} \quad \xi \to \Omega,$$

for all $\beta < \Omega$.

The following is a cheapo form of weak covering. It is crucial in what follows; it tells us that we haven't been too miserly about putting extenders on the K^c sequence.

Theorem 1.4. *Exactly one of the following holds:*

(a) $K^c \models$ *There is a Woodin cardinal,*

(b) *for* μ_0- *a.e.* $\alpha < \Omega$, $(\alpha^+)^{K^c} = \alpha^+$.

Proof. (a) $\Rightarrow$ $\neg$ (b): Every $\mathcal{J}_\beta^{K^c}$ is 1-small, so if $K^c \models \delta$ is Woodin, then $E_\gamma^{K^c} = \emptyset$ if $\gamma \ge \delta$. As Ω is measurable, there is a club class of indiscernibles for K^c, or equivalently, a countable mouse $\mathcal{N}$ which is not 1-small. Comparing $\mathcal{N}$ with K^c, we get that for μ_0- a.e. $\alpha < \Omega$, $(\alpha^+)^{K^c}$ has cofinality ω in V.

Remark. We have no use for this direction in what follows.

$\neg$ (b) $\Rightarrow$ (a):

Let $j : V \to M = \mathrm{Ult}(V, \mu_0)$ be the canonical embedding. We are assuming then

$$(\Omega^+)^{j(K^c)} < \Omega^+ .$$

(Of course, $\Omega^{+^M} = \Omega^+$.) Let $\mathcal{A} = P(\Omega) \cap j(K^c)$, so that $\mathcal{A} \in M$ and $M \models |\mathcal{A}| = \Omega$. Let E_j be the $(\Omega, j(\Omega))$ extender derived from j. By an ancient argument due to Kunen, whenever $|\mathcal{B}| = \Omega$, $E_j \cap ([j(\Omega)]^{<\omega} \times \mathcal{B}) \in M$.

(Proof: if $\mathcal{B} = \{B_\alpha \mid \alpha < \Omega\}$, then notice $\langle j(B_\alpha) \mid \alpha < \Omega\rangle = j(\langle B_\alpha \mid \alpha < \Omega\rangle) \restriction \Omega$, so $\langle j(B_\alpha) \mid \alpha < \Omega\rangle \in M$. Also, $B_\alpha \in (E_j)_c$ iff $c \in j(B_\alpha)$.)

In particular, setting

$$F = E_j \cap ([j(\Omega)]^{<\omega} \times j(K^c)),$$

we have that $F \in M$. Now it cannot be that every proper initial segment $F \restriction \nu$, $\nu < j(\Omega)$, of F belongs to $j(K^c)$, as otherwise these initial segments witness that Ω is Shelah in $j(K^c)$. But if $K^c \models$ There are no Woodins, then as $\mathcal{J}_\Omega^{j(K^c)} = K^c$, 1-smallness is not a barrier to adding these $F \restriction \nu$ to the $j(K^c)$ sequence. The following claim asserts that there are no other barriers. Its statement and proof run parallel to those of Lemma 11.4 of [FSIT].

Claim. Suppose $K^c \models$ There are no Woodin cardinals. Let $(\Omega^+)^{j(K^c)} \leq \rho < j(\Omega)$ and suppose that either $\rho = (\Omega^+)^{j(K^c)}$, or $\rho - 1$ exists and is a generator of F, or ρ is a limit of generators of F. Let G be the trivial completion of $F \restriction \rho$, and $\gamma = lhG$. Let $\boldsymbol{E}$ be the extender sequence of $j(K^c)$. Then either

(a) $\rho = (\Omega^+)^{j(K^c)}$, $\gamma \in \text{dom } \boldsymbol{E}$, and $E_\gamma = G = F \restriction \gamma$, or

(b) $\rho - 1$ exists, $\gamma \in \text{dom } \boldsymbol{E}$, and $E_\gamma = G = F \restriction \gamma$, or

(c) ρ is a limit ordinal $> (\Omega^+)^{j(K^c)}$, ρ is not a generator of F, and $\gamma \in \text{dom } \boldsymbol{E}$ and $E_\gamma = G = F \restriction \gamma$, or

(d) ρ is a limit ordinal $> (\Omega^+)^{j(K^c)}$ ρ is itself a generator of F, $\rho \notin \text{dom } \boldsymbol{E}$, and $\gamma \in \text{dom } \boldsymbol{E}$ and $E_\gamma = G$ (but $E_\gamma \neq F \restriction \gamma$), or

(e) ρ is a limit ordinal $> (\Omega^+)^{j(K^c)}$, ρ is itself a generator of F, $\rho \notin \text{dom } \boldsymbol{E}$, and $\pi(\boldsymbol{E})_\gamma = G$, where π is the canonical embedding from $J_\rho^{\boldsymbol{E}}$ to $\text{Ult}_0(J_\rho^{\boldsymbol{E}}, E_\rho)$.

Proof. By induction on ρ. As the proof is rather convoluted and follows closely the proof of Lemma 11.4 of [FSIT], we shall not give it here. The idea is as follows: we show that G satisfies, in M, all the conditions for being added to the $j(K^c)$ sequence as E_γ. We have 1-smallness by hypothesis, and coherence because F is a restriction of E_j. (If ρ is a generator of F, then we have to appeal to the condensation Theorem 8.2 of [FSIT] here.) We have the initial segment condition by induction. For our background certificates we can take (N, H), where $V_\Omega \cup \mathcal{A} \subseteq N$, $|N| = \Omega$, and $H = E_j \cap ([j(\Omega)]^{<\omega} \times N)$. (As we remarked earlier, $H \in M$.) Since G can be added to $j(K^c)$ in M as E_j, the construction of $j(K^c)$ guarantees $\gamma \in \text{dom } \boldsymbol{E}$. A "bicephalus" or "Doddage" argument guarantees $G = E_\gamma$.

The foregoing is more or less a proof in the case (a) or (c) of the conclusion holds. If (b) holds, we run into technical problems with mixed type bicephali. If (d) or (e) holds, we also run into the problem that $\mathcal{J}_\gamma^{j(K^c)}$ may not be a stage $\mathcal{M}_\xi$ of the construction of $j(K^c)$ done within M. We refer the reader to §11 of [FSIT] for a full proof.

We should note that this argument requires the iterability of the bicephali and psuedo-premice which arise. (See §11 of [FSIT] for more detail on how they arise.) We shall prove this in §9. □

We define $A_0 \subseteq \Omega$ by

$$\alpha \in A_0 \Leftrightarrow \alpha \text{ is inaccessible and } (\alpha^+)^{K^c} = \alpha^+ \text{ and}$$
$$\{\beta < \alpha \mid \beta \text{ is inaccessible and } (\beta^+)^{K^c} = \beta^+\} \text{ is not stationary in } \alpha.$$

One can easily check that if $K^c \models$ there are no Woodin cardinals, then

(i) A_0 is stationary in Ω,

(ii) A_0 has μ_0-measure 0,
(iii) $\alpha \in A_0 \Rightarrow ((\alpha^+)^{K^c} = \alpha^+ \wedge \alpha$ is inaccessible),
(iv) $\alpha \in A_0 \Rightarrow \alpha$ is not the critical point of any total-on-K^c extender from the K^c sequence.

It was in order to insure the existence of a set with the properties of A_0 that we included condition (3) in Case 1 of the construction of K^c. (Condition (2) could be dropped, but it does no harm.)

The definition of K^c which we have given in this section is unnatural in one respect: its requirement that the $\mathcal{N}_\xi$'s, and hence K^c itself, be 1-small. We believe that, were this restriction simply dropped, the resulting $\mathcal{N}'_\xi$'s would converge to a model $(K^c)'$ of height Ω, and one could show that either $(\alpha^+)^{(K^c)'} = \alpha^+$ for μ_0 a.e. $\alpha < \Omega$, or $(K^c)' \models$ there is a superstrong cardinal. What is missing at the moment is a proof that if M is a countable elementary submodel of one of the $\mathcal{N}'_\xi$'s or their associated psuedo-premice and bicephali, then M is $\omega_1 + 1$ iterable (in the sense of definition 2.8 of this paper). At the moment we can only prove this iterability result for mice which are "tame" (do not have extenders overlapping Woodin cardinals), and thus it is only to such mice that we can extend the theory presented here. (See [CMWC] and [TM].)

§2. Iterability

In this section we shall sketch a proof that if $\mathcal{T}$ is a k-maximal iteration tree on $\mathfrak{C}_k(\mathcal{N}_\xi)$, and $\mathcal{T} \upharpoonright \lambda$ is simple for all limit $\lambda < lh\mathcal{T}$, and $lh\mathcal{T}$ is a limit ordinal, then for all sufficiently large κ, $V^{\mathrm{Col}(\omega,\kappa)} \models \mathcal{T}$ has a cofinal, wellfounded branch. There are several other iterability facts we shall need, and actually we shall not prove even this one in this section, since we shall make some simplifying assumptions on $\mathcal{T}$. The reader seeking full detail and generality will find it in §9. The reader who would like to see the main ideas in our iterability proof, while avoiding full detail and generality, can content himself with this section.

In this paper, we shall diverge slightly from the terminology of [FSIT] regarding iteration trees. By an *iteration tree* we mean a system $\mathcal{T}$ obeying all the conditions required in the definition of §5 of [FSIT] except possibly the increasing length condition. That is, we do not require $\alpha < \beta \Rightarrow lh\ E_\alpha^\mathcal{T} < lh\ E_\beta^\mathcal{T}$. Iteration trees in the sense of [FSIT] we call *normal.* (We note that even in a normal tree $\mathcal{T}$, $E_\alpha^\mathcal{T}$ may not be applied to the earliest possible model. This last requirement is part of k-maximality.) Although the trees which arise in comparison processes are all normal and k-maximal for some $k \leq \omega$, we must cover more than such trees in our proof that K^c is iterable. This is because the proof of the Dodd-Jensen lemma (5.3 of [FSIT]) involves non-normal trees.

We shall say that $\mathcal{T}$ is simple iff for all sufficiently large κ, $V^{\mathrm{Col}(\omega,\kappa)} \models \mathcal{T}$ has at most one cofinal wellfounded branch. (This diverges slightly from the terminology of [FSIT].) We shall need a relative of this notion.

Definition 2.1. *Let $\mathcal{T}$ be a tree on $\mathcal{M}$ of limit length, and $\alpha \in OR$. We say $\mathcal{T}$ is α-short iff for all sufficiently large κ*

$$V^{Col(\omega,\kappa)} \models \mathcal{T} \text{ has no cofinal branch } b \text{ such that } \alpha \text{ is isomorphic to an initial segment of } OR^{\mathcal{M}_b^\mathcal{T}}.$$

(Here $\mathcal{M}_b^\mathcal{T} = \lim_{\alpha\in b} \mathcal{M}_\alpha^\mathcal{T}$.)

The next two lemmas come from the uniqueness theorem of §2 of [IT]. See also Theorem 6.1 of [FSIT]. Their formulation also owes a lot to work of Woodin, and to the Π_2^1 mouse condition of §5 of [IT].

Lemma 2.2. *Let $\mathcal{M}$ be 1-small and $\mathcal{T}$ an iteration tree on $\mathcal{M}$, and let $\lambda < lh\ \mathcal{T}$. Then for some $\alpha \in OR$, $\mathcal{T} \upharpoonright \lambda$ is α-short.*

Proof. Assume not. Let

$$\begin{aligned} \delta &= \sup\{lh\ E_\beta^\mathcal{T} \mid \beta < \lambda\}, \\ \vec{E} &= \bigcup_{\beta<\lambda} \dot{E}^{\mathcal{M}_\beta^\mathcal{T}} \upharpoonright lh\ E_\beta^\mathcal{T}, \end{aligned}$$

(so that $\vec{E}$= common value of $\dot{E}^{\mathcal{M}_b^{\mathcal{T}}} \restriction \delta$, for all cofinal, possibly generic, branches b of $\mathcal{T} \restriction \lambda$). Our hypothesis implies that $\forall\alpha \in$ OR, there are possibly generic cofinal branches $b \neq c$ of $\mathcal{T} \restriction \lambda$ such that α is in the wellfounded parts of both $\mathcal{M}_b$ and $\mathcal{M}_c$. Hence so is $L_\alpha[\vec{E}]$. By §2 of [IT] or 6.1 of [FSIT], $L[\vec{E}] \models \delta$ is Woodin. But now $\vec{E}= \dot{E}^{\mathcal{M}_\lambda^{\mathcal{T}}} \restriction \delta$, and $E_\lambda^{\mathcal{T}}$ has length $\geq \delta$. It follows that $\mathcal{M}_\lambda^{\mathcal{T}}$ is not 1-small, a contradiction. □

We do not get from 2.2 that $\mathcal{T}$ itself is α-short for some $\alpha \in$ OR. But the proof of 2.2 gives at once:

Lemma 2.3. *Let $\mathcal{T}$ be an iteration tree of limit length on a premouse $\mathcal{M}$. Then either $\mathcal{T}$ is α-short for some $\alpha \in OR$, or there is a proper class inner model with a Woodin cardinal.*

The next lemma explains the importance of these uniqueness facts in our proof of iterability. It shows that the existence of a "bad" tree on $\mathcal{M}$ reflects to the existence of a bad countable tree on a countable $\mathcal{N} \preceq \mathcal{M}$. Part (a) of the lemma is due to Woodin and the author independently; part (b) is due essentially to Woodin.

Let us use "putative iteration tree" for a system having all the properties of an iteration tree, except that its last model, if it has one, may be illfounded.

Lemma 2.4. *Let $\mathcal{T}$ be a putative iteration tree on a 1-small premouse $\mathcal{M}$ such that $\mathcal{T} \restriction \lambda$ is simple for all $\lambda < lh\ \mathcal{T}$. Suppose that either* (a) $(\mathcal{T},\mathcal{M})^\sharp$ *exists, or* (b) *There is no proper class inner model with a Woodin cardinal. Suppose also that either $\mathcal{T}$ has a last, illfounded model, or $\mathcal{T}$ has limit length and for all sufficiently large κ, $V^{Col(\omega,\kappa)} \models \mathcal{T}$ has no cofinal wellfounded branch.*

Then there is a countable $\mathcal{N} \preceq \mathcal{M}$ and a countable putative iteration tree $\mathcal{U}$ on $\mathcal{N}$ such that $\mathcal{U} \restriction \lambda$ is simple for all $\lambda < lh\ \mathcal{U}$ and either $\mathcal{U}$ has a last, illfounded model, or $\mathcal{U}$ has limit length but no cofinal wellfounded branch.

Proof. We give the proof under hypothesis (b). We also assume $lh\ \mathcal{T}$ is a limit ordinal, and leave the contrary case to the reader.

By 2.3 we can fix $\alpha \in$ OR such that for all $\lambda \leq lh\ \mathcal{T}$, $\mathcal{T} \restriction \lambda$ is α-short. Let θ be large enough that $\mathcal{T}, \mathcal{M}, \alpha \in V_\theta$ and V_θ satisfies a reasonable fragment of ZFC. Lowenheim-Skolem gives us a countable transitive H and embedding

$$\pi : H \to V_\theta$$

such that for some $(\mathcal{N},\mathcal{U},\bar{\alpha})$,

$$\pi((\mathcal{N},\mathcal{U},\bar{\alpha})) = (\mathcal{M},\mathcal{T},\alpha).$$

So $\mathcal{N} \preceq \mathcal{M}$ via $\pi \restriction \mathcal{N}$. Also, $\mathcal{U}$ is a countable iteration tree on $\mathcal{N}$ of limit length. Now $H \models \mathcal{U} \restriction \lambda$ is $\bar{\alpha}$ short, for all $\lambda \leq lh\ \mathcal{U}$. But this notion is sufficiently absolute that $\mathcal{U} \restriction \lambda$ truly is $\bar{\alpha}$ short, for all $\lambda \leq lh\ \mathcal{U}$. [Let x be a real which is a member of $H^{\mathrm{Col}(\omega,\kappa)}$, κ sufficiently large, and codes $(\mathcal{N},\mathcal{U},\bar{\alpha})$. Being $\bar{\alpha}$ short is a Π^1_1 property of x, and so absolute to $H^{\mathrm{Col}(\omega,\kappa)}$.]

Now $H \models \mathcal{U} \upharpoonright \lambda$ is simple, $\forall\lambda < lh\,\mathcal{U}$. Since $\mathcal{U} \upharpoonright \lambda$ is $\bar{\alpha}$ short and $\bar{\alpha} \in H$, this is absolute, and thus $\mathcal{U} \upharpoonright \lambda$ truly is simple for all $\lambda < lh\,\mathcal{U}$.

Similarly, if $\mathcal{U}$ has a cofinal wellfounded branch b, then $\mathrm{OR}^{\mathcal{M}_b} < \bar{\alpha}$, and thus $\mathcal{U}$ has a cofinal wellfounded branch in $H^{\mathrm{Col}(\omega,\kappa)}$ where $\kappa = (\mathrm{card}(\mathcal{N})) \cdot \mathrm{card}(\mathcal{U}) \cdot \bar{\alpha})^H$. So $\mathcal{U}$ has no cofinal wellfounded branch.

The proof under hypothesis (a) is similar. Where in case (b) we used α-shortness and Π^1_1 absoluteness, we use $(\mathcal{T},\mathcal{M})^\sharp$ and Π^1_2 absoluteness. [We have $\pi : (\mathcal{U},\mathcal{N})^{\sharp^H} \to (\mathcal{T},\mathcal{M})^\sharp$, and this guarantees $(\mathcal{U},\mathcal{N})^{\sharp^H} = (\mathcal{U},\mathcal{N})^\sharp$. That in turn implies $H[G]$ is correct for Π^1_2 statements about x, where x is a real coding $(\mathcal{U},\mathcal{N})$ in $H[G]$, and G is generic $/H$ for $\mathrm{Col}(\omega,\kappa)$, $\kappa = (\mathrm{card}\ (\mathcal{U}) \cdot \mathrm{card}(\mathcal{N}))^H$.] □

Remark. If $\kappa = \mathrm{card}(\mathcal{T}) \cdot \mathrm{card}(\mathcal{M})$, then $\forall\theta > \kappa$, $\mathcal{T}$ has a cofinal wellfounded branch in $V^{\mathrm{Col}(\omega,\theta)}$ iff $\mathcal{T}$ has a cofinal wellfounded branch in $V^{\mathrm{Col}(\omega,\kappa)}$. In particular, if $\omega = \mathrm{card}(\mathcal{T}) \cdot \mathrm{card}(\mathcal{M})$, $\mathcal{T}$ has such a branch in $V^{\mathrm{Col}(\omega,\theta)}$ iff $\mathcal{T}$ has such a branch in V.

We are ready to state the main result of this section, which concerns the iterability of countable elementary submodels of K^c and its levels. Although we could prove that such structures are iterable with respect to arbitrary trees, to do so would add a layer of notational fog to the proof for normal trees. We shall therefore prove just the iterability we need, which is iterability with respect to *linear compositions of normal trees.* We call these trees *almost normal.* More precisely, suppose $\langle\mathcal{T}_\alpha \mid \alpha < \beta\rangle$ is a sequence of normal trees such that $\mathcal{T}_0$ is on $\mathcal{M}$, $\mathcal{T}_{\alpha+1}$ is on the last model $\mathcal{M}^{\mathcal{T}_\alpha}_{\nu_\alpha}$ of $\mathcal{T}_\alpha$ for all $\alpha+1 < \beta$, and $\mathcal{T}_\lambda$ is on the direct limit of the $\mathcal{M}^{\mathcal{T}_\alpha}_{\nu_\alpha}$, for $\alpha < \lambda$, if $\lambda < \beta$ is a limit. We can form an iteration tree $\mathcal{U}$ by "laying the $\mathcal{T}_\alpha$'s end-to-end". We say $\mathcal{U}$ is generated by $\langle\mathcal{T}_\alpha \mid \alpha < \beta\rangle$, and call a tree $\mathcal{U}$ generated in this way almost normal. Such a composition $\mathcal{U}$ will generally not be maximal, even if the $\mathcal{T}_\alpha$'s are maximal, since maximality requires going back to the earliest possible model. We say $\mathcal{U}$ is *almost k-maximal* iff $\mathcal{T}_0$ is k-maximal, and $\forall\gamma < \beta$ ($\mathcal{T}_\gamma$ is j-maximal, where $j = \deg^{\mathcal{T}_\alpha}(\mathcal{M}^{\mathcal{T}_\alpha}_{\nu_\alpha})$ for all sufficiently large $\alpha < \gamma$.

Theorem 2.5. *Let $\mathcal{P} \preceq \mathfrak{C}_k(\mathcal{N}_\theta)$ for some k,θ, and $\mathcal{P}$ be countable. Let $\mathcal{T}$ be a countable, almost normal, almost k-maximal putative iteration tree on $\mathcal{P}$ such that $\mathcal{T} \upharpoonright \lambda$ is simple for all $\lambda < lh\,\mathcal{T}$. Then either $\mathcal{T}$ has successor length, and its last model is wellfounded, or $\mathcal{T}$ has limit length, and $\mathcal{T}$ has a cofinal wellfounded branch.*

Sketch of Proof. We shall give the proof in a special case which highlights the new ideas.

The simplifying assumptions we make are: $\mathcal{T}$ is normal and k-maximal, and

(1) $\mathcal{T}$ has length ω,

(2) $k = \omega$; moreover $\mathcal{P}$ is passive and $\mathcal{P} \models ZF^-$, and there is no dropping on $\mathcal{T}$,

(3) Letting $\mathcal{P}_i$ be the ith model of $\mathcal{T}$, and ν_i the sup of the generators of $E_i^{\mathcal{T}}$,

$$\mathcal{P}_i \models \quad \text{strength}(E_i^{\mathcal{T}}) \geq \nu_i\,,$$

moreover, ν_i is a limit ordinal.

As ν_i is a limit ordinal, $\mathcal{J}_\beta^{\mathcal{P}_i}$ is either type I or type III, where $\beta = lh\ E_i^{\mathcal{T}}$.

This implies that ν_i is a cardinal of $\mathcal{J}_\beta^{\mathcal{P}_i}$, and therefore, by our strength assumption in (3), ν_i is a cardinal of $\mathcal{P}_i$.

We have made assumptions (2) and (3) in part to avoid any need for "resurrection" (cf. §12 of [FSIT]) in the construction to follow.

By (2), $\rho_\omega(\mathcal{P}_i) = \text{OR}^{\mathcal{P}_i}$ for all $i \in \omega$, all ultrapowers on $\mathcal{T}$ are Σ_ω (satisfy the full Łos theorem), yet are formed using functions which belong to the model in question. It may seem that these assumptions just resurrect the "coarse structure" setting of [IT], but in fact they do not. For one thing, we don't have $\mathcal{T} \in \mathcal{P}_0$.

Because $\rho_\omega(\mathcal{P}) = \text{OR}^{\mathcal{P}}$, $\mathfrak{C}_\omega(\mathcal{N}_\theta) = \mathcal{N}_\theta$. Fix an elementary $\pi : \mathcal{P} \to \mathcal{N}_\theta$. We shall show that there is a cofinal branch b of $\mathcal{T}$ and elementary $\sigma : \mathcal{P}_b \to \mathcal{N}_\theta$ such that

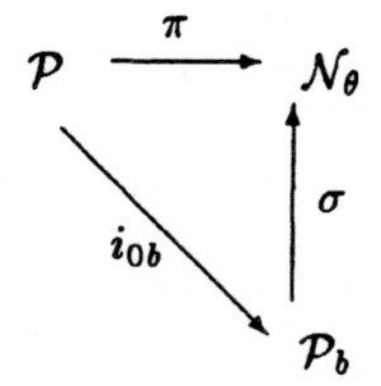

commutes.

Let $\mathcal{U}$ be the tree of attempts to build such a branch b and embedding σ. More precisely, let $\tau : \mathcal{P} \to Q$ be elementary; we shall define a tree $\mathcal{U} = \mathcal{U}(\tau, Q)$ which tries to build (b, σ) such that $\sigma : \mathcal{P}_b \to Q$ and $\tau = \sigma \circ i_{0,b}^{\mathcal{T}}$.

Fix an enumeration

$$t : \omega \overset{\text{onto}}{\to} \bigcup_{i \in \omega} (\{i\} \times \mathcal{P}_i)$$

such that $t^{-1}((e, x))$ is infinite for all (e, x) such that $x \in \mathcal{P}_e$. We then put $(\langle e_0, \ldots, e_k \rangle, \langle y_0, \ldots, y_k \rangle) \in \mathcal{U}$ iff

(a) $e_0 T e_1 T \cdots T e_k$

(b) $(\mathcal{P}_{e_k}, x_0, \ldots, x_k) \equiv (Q, y_0, \ldots, y_k)$, where $\forall\ i \leq k$

(i) $t(i) = (e, x)$ for $e \in [0, e_i]_T$ implies $x_i = i_{e,e_k}^{\mathcal{T}}(x)$,
(ii) $t(i) = (e, x)$ for $e \notin [0, e_i]_T$ implies $x_i = \emptyset$,
(iii) $t(i) = (0, x)$ implies $y_i = \tau(x)$.

Remark. $\langle x_0, \ldots, x_k \rangle$ in (b) is determined by $\langle e_0, \ldots, e_k \rangle$.

Let $\mathcal{U} = \mathcal{U}(\tau, Q)$ and $i \in \omega$, and suppose we are given an embedding $\sigma : \mathcal{P}_i \to Q$ such that $\tau = \sigma \circ i^{\mathcal{T}}_{0,i}$. From σ we get an initial segment of a branch of $\mathcal{U}(\tau, Q)$; let

$$p(i, \sigma, \tau, Q) = (\langle e_0, \ldots, e_k \rangle, \langle y_0, \ldots, y_k \rangle)$$

where $\langle e_0, \ldots, e_k \rangle$ is the increasing enumeration of $[0, i]_T$ and letting $\langle x_0, \ldots, x_k \rangle$ come from $\langle e_0, \ldots, e_k \rangle$ as in the definition of $\mathcal{U}(\tau, Q)$, we have $y_j = \sigma(x_j)$ for all $j \leq k$.

Theorem 2.5 is proved if we show that $\mathcal{U}(\pi, \mathcal{N}_\theta)$ has an infinite branch. Let us assume otherwise toward a contradiction.

By "coarse premouse" we mean a premouse in the sense of [IT]; that is, a structure $\mathcal{M} = (M, \in, \delta)$, where M is transitive which is power-admissible and satisfies choice, the full collection schema for domains $\subseteq V^{\mathcal{M}}_\delta$, and the full separation schema. We also require that ${}^\omega M \subseteq M$, and that δ be inaccessible in M. Write $\delta^{\mathcal{M}} = \delta$.

Let

$$\mathbb{C} = \langle \mathcal{N}_\xi \mid \xi < \gamma \rangle \qquad (\gamma \leq \Omega)$$

be the construction done in §1. Notice that $\mathbb{C}$ is definable from no parameters over V_Ω. (Here γ is the first place $< \Omega$ where the construction breaks down, if any, and $\gamma = \Omega$ otherwise. Thus $\theta < \gamma$.) It follows that if $\mathcal{R} = (R, \in, \delta)$ is any coarse premouse, then $\mathbb{C}^{\mathcal{R}}$ makes sense: we interpret the definition of $\mathbb{C}$ inside $V^{\mathcal{R}}_\delta$.

If $\mathcal{R}$ is a coarse premouse, then a *cutoff point* of $\mathcal{R}$ is an ordinal ξ such that $\delta^{\mathcal{R}} < \xi < \mathrm{OR}^{\mathcal{R}}$ and $(V^{\mathcal{R}}_\xi, \in, \delta^{\mathcal{R}})$ is a coarse premouse.

We now define by induction on i triples $(\pi_i, Q_i, \mathcal{R}_i)$ with the following properties:

(1) $\mathcal{R}_i$ is a coarse premouse,

(2) Q_i is an "N-model" of the construction $\mathbb{C}^{\mathcal{R}_i}$, moreover $\pi_i : \mathcal{P}_i \to Q_i$ elementarily,

(3) for all $j < i$, $\mathcal{R}_j$ agrees with $\mathcal{R}_i$, through $\pi_j(\nu_j)$,

(Recall that $\nu_j = \nu(E^{\mathcal{T}}_j)$ = strict sup of the generators of $E^{\mathcal{T}}_j$. We say coarse premice $\mathcal{R}$ and $\mathcal{S}$ agree through η iff $V^{\mathcal{R}}_\eta = V^{\mathcal{S}}_\eta$.)

(4) for all $j < i$, $\pi_i \restriction \nu_j = \pi_j \restriction \nu_j$; moreover Q_j agrees with Q_i through $\pi_j(\nu_j)$,

(Ordinary, "fine" premice Q and $\mathcal{R}$ agree through η iff $\mathcal{J}^Q_\eta = \mathcal{J}^{\mathcal{R}}_\eta$.)

(5) Let $\mathcal{U} = \mathcal{U}(\pi_i \circ i^{\mathcal{T}}_{0,i}, Q_i)$ and

$$p = p(i, \pi_i, \pi_i \circ i^{\mathcal{T}}_{0,i}, Q_i)\,.$$

(So $\mathcal{U}$ is a tree in $\mathcal{R}_i$ and p is a node of $\mathcal{U}$.) Then $\mathcal{U}$ is wellfounded, and the order type of the set of cutoff points of $\mathcal{R}_i$ is at least $|p|_{\mathcal{U}}$,

(6) if $i > 0$, then $\mathcal{R}_i \in \mathcal{R}_{i-1}$.

Clause (6) gives us the desired contradiction.

Base step: Set

$$\begin{aligned}\pi_0 &= \pi\,,\\ Q_0 &= \mathcal{N}_\theta\,.\end{aligned}$$

By assumption, $\mathcal{U} = \mathcal{U}(\pi_0, Q_0)$ is wellfounded. Set

$$\mathcal{R}_0 = (V_\xi, \in, \Omega)$$

where ξ is the $|\mathcal{U}|$th ordinal $\alpha > \Omega$ such that $(V_\alpha, \in, \Omega)$ is a coarse premouse. Our applicable induction hypotheses, namely (1), (2), and (5), clearly hold.

Inductive step. We are given $\langle(\pi_j, Q_j, \mathcal{R}_j) \mid j \leq i\rangle$. Let $j = T$-pred $(i+1)$, and set

$$Q'_{i+1} = \mathrm{Ult}_\omega(Q_j, \pi_i(E_i^{\mathcal{T}}))\,.$$

$Q_j \models ZF^-$, so the ultrapower is formed using functions belonging to Q_j.

Notice that the ultrapower makes sense. For set $\bar\kappa =$ crit $E_i^{\mathcal{T}}$ and $\kappa = \pi_i(\bar\kappa)$. Let $E = \pi_i(E_i^{\mathcal{T}})$. The rules for iteration trees guarantee $\bar\kappa < \nu_j$. Induction hypothesis (4) states that $\pi_i \restriction \nu_j = \pi_j \restriction \nu_j$; thus $\kappa < \pi_j(\nu_j)$. But $\pi_j(\nu_j)$ is a cardinal of Q_j, and Q_i agrees with Q_j through $\pi_j(\nu_j)$. Thus $P(\kappa)^{Q_j} \subseteq Q_i$, and the ultrapower makes sense. (We may have subsets of κ in Q_i but not Q_j; $\pi_j(\nu_j)$ may not be a cardinal of Q_i. So E may measure more sets than necessary.)

Let $\sigma : \mathcal{P}_{i+1} \to Q'_{i+1}$ be given by the shift lemma:

$$\sigma([a, f]^{\mathcal{P}_j}_{E_i^{\mathcal{T}}}) = [\pi_i(a), \pi_j(f)]^{Q_j}_E\,.$$

We have that σ is well defined and elementary, that Q'_{i+1} agrees with Q_i through all $\eta < lh\ E$, that $\sigma \restriction lh\ E_i^{\mathcal{T}} = \pi_i \restriction lh\ E_i^{\mathcal{T}}$, and that $\sigma \circ i^{\mathcal{T}}_{j,i+1} = i_E \circ \pi_j$ where $i_E : Q_j \to Q'_{i+1}$ is the canonical embedding.

The following little lemma will be useful.

Lemma 2.6. *Suppose $\mathcal{J}_\beta^{\mathcal{N}_\eta}$ is an initial segment of $\mathcal{N}_\eta$ such that*

$$\forall\kappa < \omega\beta[(\mathcal{N}_\eta \models \kappa \quad \textit{is a cardinal}) \Leftrightarrow (\mathcal{J}_\beta^{\mathcal{N}_\eta} \models \kappa \quad \textit{is a cardinal})]\,;$$

then there is a $\xi \leq \eta$ such that $\mathcal{J}_\beta^{\mathcal{N}_\eta} = \mathcal{N}_\xi$.

Proof. We may assume $\omega\beta < \mathrm{OR} \cap \mathcal{N}_\eta$.

Let $\xi \leq \eta$ be least such that $\mathcal{J}_\beta^{\mathcal{N}_\eta}$ is an initial segment of $\mathcal{N}_\xi$. Clearly, if ξ is a limit then $\mathcal{J}_\beta^{\mathcal{N}_\eta} = \mathcal{N}_\xi$, and we are done. Therefore we suppose $\xi = \tau + 1$. We also suppose $\mathcal{J}_\beta^{\mathcal{N}_\eta} \neq \mathcal{N}_{\tau+1}$, and this implies that $\mathcal{J}_\beta^{\mathcal{N}_\eta}$ is an initial segment of $\mathcal{M}_\tau$. Since $\mathcal{J}_\beta^{\mathcal{N}_\eta}$ is not an initial segment of $\mathcal{N}_\tau$, and $\mathcal{M}_\tau = \mathfrak{C}_\omega(\mathcal{N}_\tau)$, we have from the proof of Theorem 8.1 of [FSIT] that $\rho_\omega(\mathcal{N}_\tau) < \mathrm{OR}^{\mathcal{N}_\tau}$ and $\mathcal{N}_\tau \models \rho_\omega(\mathcal{N}_\tau)^+$ exists, moreover $\omega\beta$ is strictly larger than $(\rho_\omega(\mathcal{N}_\tau)^+)^{\mathcal{N}_\tau}$. Now let $\delta = \inf\{\rho_\omega(\mathcal{N}_\theta) \mid \tau \leq \theta < \eta\}$, so that $\delta \leq \rho_\omega(\mathcal{N}_\tau)$ and $(\delta^+)^{\mathcal{N}_\tau} \leq (\rho_\omega(\mathcal{N}_\tau)^+)^{\mathcal{N}_\tau} < \omega\beta$. Then $(\delta^+)^{\mathcal{N}_\tau}$ is a cardinal of $\mathcal{M}_\tau$ (by §8 of [FSIT]), and thus it is a cardinal of $\mathcal{J}_\beta^{\mathcal{N}_\eta}$. On the other hand, the defining property of δ guarantees $(\delta^+)^{\mathcal{N}_\tau}$ is not a cardinal of $\mathcal{N}_\eta$. This contradicts the hypothesis of 2.6. □

Since E is on the Q_i sequence, $E = E_\beta^{Q_i}$ where $\beta = lh\ E$. Let $Q_i = \mathcal{N}_\eta^{\mathcal{R}_i}$. Lemma 2.6 then gives us a $\xi \leq \eta$ such that $\mathcal{J}_\beta^{Q_i} = \mathcal{N}_\xi^{\mathcal{R}_i}$. (We apply 2.6 within $\mathcal{R}_i$. Our simplifying assumptions tell us that $\pi_i(\nu_i)$ is the largest cardinal of $\mathcal{J}_\beta^{\mathcal{N}_\eta}$, and $\pi_i(\nu_i)$ remains a cardinal in $\mathcal{N}_\eta$. Thus 2.6 applies.)

Remark. Without our simplifying assumptions we don't get that $\pi_i(\nu_i)$ is a cardinal of Q_i, and therefore there may be no ξ such that $\mathcal{J}_\beta^{Q_i} = \mathcal{N}_\xi^{\mathcal{R}_i}$. At this point in the general argument we need to resurrect a background extender for E by inverting certain collapses.

If we were in the situation of [FSIT] we would now have a "background extender" F for E such that $\mathrm{Ult}(\mathcal{R}_j, F)$ makes sense. We would let $\mathcal{R}'_{i+1} = \mathrm{Ult}(\mathcal{R}_j, F)$, and then take $\mathcal{R}_{i+1}$ to be the collapse of a suitable hull of a suitable cutoff point of $\mathcal{R}'_{i+1}$. Q_{i+1} would be the image of Q'_{i+1} under collapse. Now, however, we have no such F (after all, $V_{\kappa+1}^{\mathcal{R}_j} = V_{\kappa+1}^{\mathcal{R}_i}$, so F would be a full extender in $\mathcal{R}_i$). So instead we get a suitable background certificate (N, F) for E in $\mathcal{R}_i$. Since N is large enough, and in particular $V_\kappa^{\mathcal{R}_j} = V_\kappa^{\mathcal{R}_i} \subseteq N$, we can take an analogue of the hull producing $\mathcal{R}_{i+1}$ and Q_{i+1} "almost everywhere" below κ. We get $\mathcal{R}_{i+1}(\bar{u})$, $Q_{i+1}(\bar{u})$ for F_b are $\bar{u}$, for a suitable b. We then let $\mathcal{R}_{i+1} = [b, \lambda\bar{u}\mathcal{R}_{i+1}(\bar{u})]_F^N$ and $Q_{i+1} = [b, \lambda\bar{u}Q_{i+1}(\bar{u})]_F^N$.

Let

$$\mathcal{A} = \pi_i'' \bigcup_{n<\omega} P([\bar{\kappa}]^n)^{\mathcal{P}_i} = \pi_j'' \bigcup_{n<\omega} P([\bar{\kappa}]^n)^{\mathcal{P}_j}.$$

Since $\mathcal{A}$ is a countable subset of $\mathcal{R}_i$, $\mathcal{A} \in \mathcal{R}_i$ and is countable in $\mathcal{R}_i$. Also $\mathcal{A} \subseteq \bigcup_{n<\omega} P([\kappa]^n)^{\mathcal{N}_\xi^{\mathcal{R}_i}}$, where $\mathcal{N}_\xi^{\mathcal{R}_i} = \mathcal{J}_\beta^{Q_i}$. E is the last extender of $\mathcal{N}_\xi^{\mathcal{R}_i}$, so we can let

$$\mathcal{R}_i \models (N, F) \text{ is an } \mathcal{A}\text{-certificate for } \mathcal{N}_\xi^{\mathcal{R}_i}.$$

Since ${}^\omega\mathrm{Ult}(N, F) \subseteq \mathrm{Ult}(N, F)$, $\pi_i \restriction \nu_i \in \mathrm{Ult}(N, F)$. Let us pick an F support b and functions $\lambda\bar{u} \cdot \pi_i(\bar{u})$, $\lambda\bar{u} \cdot \nu(\bar{u})$ such that

$$\begin{aligned} \pi_i \restriction \nu_i &= [b, \lambda\bar{u} \cdot \pi_i(\bar{u})]_F^N, \\ \pi_i(\nu_i) &= [b, \lambda\bar{u} \cdot \nu(\bar{u})]_F^N. \end{aligned}$$

We may assume that $\pi_i(\bar{u}), \nu(\bar{u}) \in V_\kappa^N$ for all $\bar{u}$.

Claim. For F_b a.e. $\bar{u}$, there are in V_κ^N: a coarse premouse $\mathcal{R}$, an "$\mathcal{N}$ model" Q of $\mathbb{C}^{\mathcal{R}}$, and an elementary embedding $\pi : \mathcal{P}_{i+1} \to Q$ such that

(1) $V_{\nu(\bar{u})}^{\mathcal{R}} = V_{\nu(\bar{u})}^N$, and $\mathcal{J}_{\nu(\bar{u})}^Q = \mathcal{J}_{\nu(\bar{u})}^{Q_i}$

(2) $\pi \restriction \nu_i = \pi_i(\bar{u})$,

and

(3) letting $\mathcal{U} = \mathcal{U}(\pi \circ i^{\mathcal{T}}_{0,i+1}, Q)$ and $p = p(i+1, \pi, \pi \circ i^{\mathcal{T}}_{0,i+1}, Q)$, we have: $\mathcal{U}$ is wellfounded, and there are in order type at least $|p|_{\mathcal{U}}$ cutoff points of $\mathcal{R}$.

Proof. F_b measures the set of such $\bar{u}$, as the quantifiers in its definition range over V^N_κ, and $\mathcal{J}^{Q_i}_\kappa \in N$. Let $\bar{u} \in X$ if and only if $\bar{u} \in [\kappa]^{|b|}$ and the claim fails for $\bar{u}$, and suppose toward a contradiction that $X \in F_b$.

Fix an enumeration $\mathcal{P}_{i+1} = \{x_n \mid n < \omega\}$ of $\mathcal{P}_{i+1}$. For $n < \omega$ let

$$\sigma(x_n) = [c_n, f_n]^{Q_j}_E,$$

where $c_n \in [\pi_i(\nu_i)]^{<\omega}$ and $f_n \in Q_j$. If $x_n < \nu_i$, so that $\sigma(x_n) = \pi_i(x_n)$, then we choose $c_n = \{\pi_i(x_n)\}$ and $f_n =$ identity function.

Subclaim A. There is a set $Y_n \in F_{c_0 \cup \cdots \cup c_n}$ such that if $t : \bigcup_{i \le n} c_i \to \kappa$ is order preserving, and $t'' \bigcup_{i \le n} c_i \in Y_n$, then

$$(\mathcal{P}_{i+1}, x_0, \ldots, x_n) \equiv (Q_j, f_0(t''c_0) \cdots f_n(t''c_n)).$$

Proof. Note $(\mathcal{P}_{i+1}, x_0 \cdots x_n) \equiv (Q'_{i+1}, \sigma(x_0), \ldots, \sigma(x_n))$. Now let $Q'_{i+1} \models \varphi[\sigma(x_0), \ldots, \sigma(x_n)]$. By Los' theorem there is a set $Y^\varphi_n \in E_{c_0 \cup \cdots \cup c_n}$ such that if $t : \bigcup_{i \le n} c_i \to \kappa$ is order-preserving and $t'' \bigcup_{i \le n} c_i \in Y^\varphi_n$, then $Q_j \models \varphi[f_0(t''c_0), \ldots, f_n(t''c_n)]$. We can choose $Y^\varphi_n \in \operatorname{ran} \pi_j$ because $\{f_0, \ldots, f_n\} \subseteq \operatorname{ran} \pi_j$. Thus $Y^\varphi_n \in \mathcal{A}$, and hence $Y^\varphi_n \in F_{c_0 \cup \cdots \cup c_n}$. Let $Y_n = \bigcap_\varphi Y^\varphi_n$; this works because $F_{c_0 \cup \cdots \cup c_n}$ is countably complete. □

Subclaim B. If $x_n < \nu_i$, then there is a set $Z_n \in F_{c_n \cup b}$ such that if $t : c_n \cup b \to \kappa$ is order preserving and $t''(c_n \cup b) \in Z_n$, then $\pi_i(t''b)(x_n) = f_n(t''c_n) < \nu(t''b)$.

Proof.

$$\operatorname{Ult}(N, F) \models [b, \lambda \bar{u} \cdot \pi_i(\bar{u})]^N_F(x_n) = [c_n, f_n]^N_F < [b, \lambda \bar{u} \cdot \nu(\bar{u})]^N_F$$

(noting that $[c_n, f_n]^N_F = [\{\pi_i(x_n)\}, \text{identity}]^N_F = \pi_i(x_n)$ since $x_n < \nu_i$). The subclaim now follows from Los' theorem for $\operatorname{Ult}(N, F)$. □

Subclaim C. If $x_n = i^{\mathcal{T}}_{j,i+1}(y)$, then there is a set $W_n \in F_{c_n}$ such that whenever $t''c_n \in W_n$, $f_n(t''c_n) = \pi_j(y)$.

Proof.

$$\begin{aligned} [c_n, f_n]^{Q_j}_E &= \sigma(i^{\mathcal{T}}_{j,i+1}(y)) \\ &= i_E(\pi_j(y)) \\ &= [c_n, \lambda \bar{u} \cdot \pi_j(y)]^{Q_j}_E . \end{aligned}$$

By Los' theorem for $\operatorname{Ult}(Q_j, E)$, there is a set $W_n \in E_{c_n}$ as desired. But we can assume $W_n \in \operatorname{ran} \pi_j$, so that $W_n \in \mathcal{A}$, and thus $W_n \in F_{c_n}$. □

Since F is countably complete, we can find

$$t : \bigcup_{n<\omega} c_n \cup b \to \kappa$$

order preserving such that $t''b \in X$, and $t'' \bigcup_{i \leq n} c_i \in Y_n$, and $t''(c_n \cup b) \in Z_n$, and $t''c_n \in W_n$, for all $n \leq \omega$. Letting

$$\psi(x_n) = f_n(t''c_n),$$

$\psi : \mathcal{P}_{i+1} \to Q_j$ elementarily, and $\psi \restriction \nu_i = \pi_i(t''b)$, and $\pi_j = \psi \circ i^{\mathcal{T}}_{j,i+1}$. Let

$$\mathcal{U} = \mathcal{U}(\pi_j \circ i^{\mathcal{T}}_{0,j}, Q_j),$$

and

$$p = p(j, \pi_j, \pi_j \circ i^{\mathcal{T}}_{0,j}, Q_j),$$

so that by induction $\mathcal{R}_j$ has at least $|p|_{\mathcal{U}}$ many cutoff points. Let

$$q = p(i+1, \psi, \pi_j \circ i^{\mathcal{T}}_{0,j}, Q_j).$$

One can check easily that q is a proper extension of p in $\mathcal{U}$; this is true because $jTi+1$ and $\pi_j = \psi \circ i^{\mathcal{T}}_{j,i+1}$. Thus there is an $\eta \in \mathrm{OR}^{\mathcal{R}_j}$ such that

$$\eta = |q|_{\mathcal{U}}\text{th} \quad \text{cutoff point of} \quad \mathcal{R}_j.$$

Set

$$\begin{aligned} \mathcal{R} &= \text{transitive collapse of closure of } V^{\mathcal{R}_j}_{\nu(t''b)} \cup \{Q_j\} \\ &\quad \text{under Skolem functions for } V^{\mathcal{R}_j}_{\eta} \text{ and } \omega\text{-sequences} \\ Q &= \text{image of } Q_j \text{ under collapse,} \\ \pi &= \text{image of } \psi \text{ under collapse.} \end{aligned}$$

(Since ${}^{\omega}\mathcal{R} \subseteq \mathcal{R}$, $\mathcal{T}$ and ψ belong to the uncollapsed hull. So also do $\mathcal{U}$, p, and q.)

Clearly $(Q, \mathcal{R}, \pi) \in V^{\mathcal{R}_j}_{\kappa} = V^N_{\kappa}$. Moreover, $(Q, \mathcal{R}, \pi)$ witnesses the truth of the claim for $\bar{u} = t''b \in X$.

For (2): $\psi \restriction \nu_i = \pi \restriction \nu_i$ because we put all of $\nu(t''b)$ into the hull collapsing to $\mathcal{R}$, and $\psi(x_n) < \nu(t''b)$ for $x_n < \nu_i$ by B. But $\psi \restriction \nu_i = \pi_i(t''b)$ by B, as we observed earlier.

For (3): Since $\pi_j = \psi \circ i^{\mathcal{T}}_{j,i+1}$, $\pi_j \circ i^{\mathcal{T}}_{0,j} = \psi \circ i^{\mathcal{T}}_{0,i+1}$, and

$$\begin{aligned} \mathcal{U} &= \mathcal{U}(\psi \circ i^{\mathcal{T}}_{0,i+1}, Q_j), \\ q &= p(i+1, \psi, \psi \circ i^{\mathcal{T}}_{0,i+1}, Q_j); \end{aligned}$$

moreover $V^{\mathcal{R}_j}_{\eta}$ has at least $|q|_{\mathcal{U}}$ cutoff points. This is first order, so $\mathcal{R}$ has at least $|\bar{q}|_{\bar{\mathcal{U}}}$ cutoff points, where $\bar{q}$ = collapse of q and $\bar{\mathcal{U}}$ = collapse of $\mathcal{U}$. Since $\bar{\mathcal{U}} = \mathcal{U}(\pi \circ i^{\mathcal{T}}_{0,i+1}, Q)$ and $\bar{q} = p(i+1, \pi, \pi \circ i^{\mathcal{T}}_{0,i+1}, Q)$, we are done.

This proves the claim. □

By AC in N, we have a function

$$\bar{u} \mapsto (\mathcal{R}(\bar{u}), \mathcal{Q}(\bar{u}), \pi(\bar{u}))$$

defined F_b a.e., said function in N, picking witnesses to the claim. Let

$$\begin{aligned}\mathcal{R}_{i+1} &= [b, \lambda\bar{u}\cdot\mathcal{R}(\bar{u})]_F^N ,\\ \mathcal{Q}_{i+1} &= [b, \lambda\bar{u}\cdot\mathcal{Q}(\bar{u})]_F^N ,\end{aligned}$$

and

$$\pi_{i+1} = [b, \lambda\bar{u}\cdot\pi(\bar{u})]_F^N .$$

By (1) of the claim, $V^{\mathcal{R}_{i+1}}_{\pi_i(\nu_i)} = V^{\mathrm{Ult}(N,F)}_{\pi_i(\nu_i)} = V^{\mathcal{R}_i}_{\pi_i(\nu_i)}$. So $\mathcal{R}_{i+1}$ agrees with $\mathcal{R}_k$, $k \leq i$, as desired. By (2), $\pi_{i+1} \restriction \nu_i = \pi_i \restriction \nu_i$. To show that $\mathcal{Q}_{i+1}$ agrees with $\mathcal{Q}_i$ below $\pi_i(\nu_i)$, we argue as follows. Let $\gamma < \pi_i(\nu_i)$. Since $\mathcal{Q}_i$ agrees with $\mathcal{Q}(\bar{u})$ below $\nu(\bar{u})$, for F_b a.e. $\bar{u}$, we easily get

$$\mathcal{J}_\gamma^{\mathcal{Q}_{i+1}} = [\{\gamma\}, f]_F^N , \text{ where } f(\alpha) = \mathcal{J}_\alpha^{\mathcal{Q}_i} \text{ for all } \alpha < \kappa .$$

Now clearly $f \in \mathcal{Q}_i$, and the coherence condition on E, which is on the $\mathcal{Q}_i$ sequence, gives

$$\mathcal{J}_\gamma^{\mathcal{Q}_i} = [\{\gamma\}, f]_E^{\mathcal{J}_\beta^{\mathcal{Q}_i}} .$$

But then the agreement between E and F guarantees $\mathcal{J}_\gamma^{\mathcal{Q}_{i+1}} = \mathcal{J}_\gamma^{\mathcal{Q}_i}$. (It is enough to see that if $\beta_0 < \beta_1 < \omega\gamma$ and $\varphi(v_0, v_1)$ is a formula, then $\mathcal{J}_\gamma^{\mathcal{Q}_{i+1}} \models \varphi[\beta_0, \beta_1]$ iff $\mathcal{J}_\gamma^{\mathcal{Q}_i} \models \varphi[\beta_0, \beta_1]$, because there is a uniformly definable surjection of $\omega\gamma$ onto $\mathcal{J}_\gamma^{\mathcal{Q}}$. Let us assume $\beta_0 < \beta_1 < \gamma$ and $\mathcal{J}_\gamma^{\mathcal{Q}_{i+1}} \models \varphi[\beta_0, \beta_1]$. Then for $F_{\{\beta_0,\beta_1,\gamma\}}$ a.e. $\bar{u}$, $\mathcal{J}_{u_2}^{\mathcal{Q}_i} \models \varphi[u_0, u_1]$. The set of such $\bar{u}$ is in $\mathcal{A}$, and so is measured the same way by $E_{\{\beta_0,\beta_1,\gamma\}}$.)

The remaining induction hypotheses for our construction are easy to check. This completes our proof of 2.5 under the simplifying assumptions (1)-(3) above.

We now sketch how to do without our first simplifying assumption, that $lh\ \mathcal{T} = \omega$.

We call a sequence $\langle(\pi_j, \mathcal{Q}_j, \mathcal{R}_j) \mid j \leq i\rangle$ satisfying our inductive hypotheses (1)-(4) an *enlargement* of $\mathcal{T}$. Suppose we are given a simple $\mathcal{T}$ of length $\omega + 1$ and want to construct an enlargement $\langle(\pi_j, \mathcal{Q}_j, \mathcal{R}_j) \mid j \leq i\rangle$ of $\mathcal{T}$. Let us assume that our simplifying assumptions (2) and (3) hold of $\mathcal{T}$. Let $\mathcal{P}_j$ be the jth model of $\mathcal{T}$. We are given by hypothesis

$$\pi_0 : \mathcal{P}_0 \to \mathcal{Q}_0$$

where

$$\mathcal{Q}_0 = \mathcal{N}_\theta , \quad \text{the} \quad \theta\text{th model of} \quad \mathbb{C}^V$$

and π_0 is elementary. For any $\tau : \mathcal{P} \to \mathcal{Q}$ let

$$\mathcal{U}(\tau, Q) \quad = \quad \text{tree of attempts to build a pair } (c,\sigma), \text{ where } c \text{ is a cofinal branch of } \mathcal{T} \restriction \omega,\ c \neq [0,\omega]_T, \text{ and } \tau = \sigma \circ i^{\mathcal{T}}_{0c}\,.$$

Note because $\mathcal{T}$ is simple, $\mathcal{U}(\tau, Q)$ is wellfounded for all τ, Q. For $j \in \omega$ such that $j \notin [0,\omega]_T$, and $\psi : \mathcal{P}_j \to Q$ such that

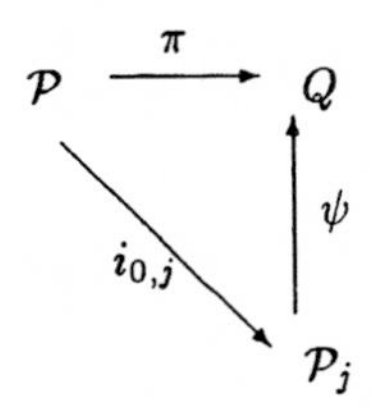

commutes, set

$$p(j, \psi, \tau, Q) \quad = \quad \text{canonical initial segment of length } |\{k \mid kTj \& \neg kT\omega\}| \text{ of a branch of } \mathcal{U}(\tau, Q) \text{ given by } (j, \psi)\,.$$

We define by induction on $i \leq \omega$ enlargements $\mathcal{E}^i$ of $\mathcal{T} \restriction i+1$. There are two cases.

Case 1. $i + 1 \notin [0,\omega]_T$.

In this case we proceed exactly as we did in the construction given in the length ω case. Let

$$\mathcal{E}^i = \langle (\pi_k, Q_k, \mathcal{R}_k) \mid k \leq i \rangle\,.$$

Let $j = T\text{-pred}(i+1)$. Let $\mathcal{U} = \mathcal{U}(\pi_j \circ i^{\mathcal{T}}_{0,j}, Q_j)$ and $p = p(j, \pi_j, \pi_j \circ i^{\mathcal{T}}_{0j}, Q_j)$. Our inductive hypotheses guarantee that $\mathcal{R}_j$ has $|p|_{\mathcal{U}}$ many cutoff points. Arguing as before, we get a background extender F for $\pi_i(E^{\mathcal{T}}_i)$, and for F_b a.e. $\bar{u}$ an embedding $\psi : \mathcal{P}_{i+1} \to Q_j$ such that $\pi_j = \psi \circ i^{\mathcal{T}}_{j,i+1}$ and $\psi \restriction \nu_i = \pi_i(\bar{u}) \restriction \nu_i$. (Here $[b, \lambda\bar{u} \cdot \pi_i(\bar{u})]^N_F = \pi_i \restriction \nu_i$.) We then set $q = p(i+1, \psi, \pi_j \circ i^{\mathcal{T}}_{0,j}, Q_j)$.)

Since $i + 1 \notin [0,\omega]_T$, q extends p in $\mathcal{U}$. Thus, for F_b a.e. $\bar{u}$, we are given a cutoff point of $\mathcal{R}_j$ at which to take a hull.

As before, we do this and collapse, producing $\mathcal{R}(\bar{u})$, $Q(\bar{u})$, $\pi(\bar{u})$. We then take $\mathcal{R}_{i+1} = [b, \lambda\bar{u}\mathcal{R}(\bar{u})]^N_F$, $Q_{i+1} = [b, \lambda\bar{u}Q(\bar{u})]^N_F$, $\pi_{i+1} = [b, \lambda\bar{u} \cdot \pi_i(\bar{u})]^N_F$. Set $\mathcal{E}^{i+1} = \mathcal{E}^i{}^\frown\langle \pi_{i+1}, Q_{i+1}, \mathcal{R}_{i+1} \rangle$.

Case 2. $i + 1 \in [0,\omega]_T$.

Again, let $\mathcal{E}^i = \langle (\pi_j, Q_j, \mathcal{R}_j) \mid j \leq i \rangle$, and let $j = T\text{-pred}(i+1)$. Arguing as before, we get "measure one many"

$$\psi : \mathcal{P}_{i+1} \to Q_j$$

such that $\pi_j = \psi \circ i^{\mathcal{T}}_{j,i+1}$. In this case, we are not given an ordinal at which to take a hull; we're on the wellfounded branch of $\mathcal{T}$ and so don't expect to get such ordinals. Along this branch, we shall realize the models of $\mathcal{T}$ back in V; that is, we take

$$\begin{aligned} \mathcal{R}_{i+1} &= \mathcal{R}_j , \\ Q_{i+1} &= Q_j , \\ \pi_{i+1} &= \psi , \end{aligned}$$

for a ψ chosen to meet certain "measure one" conditions. (Thus by induction, $\mathcal{R}_{i+1} = (V_\xi, \in, \Omega), Q_{i+1} = \mathcal{N}_\theta$, and $\pi_{i+1} \circ i^{\mathcal{T}}_{0,i+1} = \pi_0$.)

In order to do this, we must redefine $(\pi_k, Q_k, \mathcal{R}_k)$ for $j \leq k \leq i$, as otherwise our inductive hypotheses on agreement will fail. (After all, if $\bar{\kappa} =$ crit $E^{\mathcal{T}}_i$, then $\pi_j(\bar{\kappa}) = \psi \circ i^{\mathcal{T}}_{j,i+1}(\bar{\kappa}) > \psi(\bar{\kappa})$.)

First, we find new $(\pi'_k, Q'_k, \mathcal{R}'_k)$ for $j \leq k \leq i$ such that $(\pi'_k, Q'_k, \mathcal{R}'_k) \in \mathrm{Ult}(N, F)$. (As in case 1, (N, F) is a ran π_i-certificate for $\mathcal{J}^{Q_i}_\beta$ where $\beta = lh(\pi_i(E^{\mathcal{T}}_i))$.) For this, we must suppose that our induction hypothesis on the number of cutoff points gives us, for each k s.t. $j \leq k \leq i$, a cutoff point η_k of $\mathcal{R}_k$ which we can now afford to drop to. Let G be the finite set of relevant parameters, and

$$\begin{aligned} \mathcal{R}'_k &= \text{collapse of Skolem closure of } G \cup V^{\mathcal{R}_k}_{\pi_k(\nu_k)} \text{ inside } V^{\mathcal{R}_k}_{\eta_k} , \\ Q'_k &= \text{image of } \; Q_k \; \text{ under collapse}, \\ \pi'_k &= \text{collapse} \circ \pi_k , \end{aligned}$$

Now $(\pi'_k, Q'_k, \mathcal{R}'_k)$ is coded by a subset of $V^{\mathcal{R}_k}_{\pi_k(\nu_k)}$ belonging to $\mathcal{R}_k$. Let us add to our inductive agreement hypotheses

$$V^{\mathcal{R}_k}_{\pi_k(\nu_k)+1} \subseteq \mathcal{R}_i \qquad (k \leq i) .$$

(As our background extenders are "$\nu + 1$" strong, this is consistent with the construction in Case 1.) It follows that $V^{\mathcal{R}_k}_{\pi_k(\nu_k)+1} \subseteq \mathrm{Ult}(N, F)$, where (N, F) is the background certificate in $\mathcal{R}_i$ for $\pi_i(E^{\mathcal{T}}_i)$. Let for $j \leq k \leq i$,

$$(\pi'_k, Q'_k, \mathcal{R}'_k) = [b, \lambda \bar{u} \cdot (\pi'_k(\bar{u}), Q'_k(\bar{u}), \mathcal{R}'_k(\bar{u}))]^N_F .$$

Then using $\langle (\pi'_k(\bar{u}), \mathcal{R}'_k(\bar{u})) \mid j \leq k \leq i \rangle$ in the same way that we used $\pi_i(\bar{u})$ in Case 1, we can define additional measure one sets for F so that by meeting them we guarantee that

$$\mathcal{E}^{i+1} = \mathcal{E}^i \restriction j^\frown \langle (\pi'_k(\bar{u}), Q'_k(\bar{u}), \mathcal{R}'_k(\bar{u})) \mid j \leq k \leq i \rangle^\frown (\psi, Q_j, \mathcal{R}_j)$$

is an enlargement with the desired properties. (Notice that if $k < j$, then $\nu_k < \bar{\kappa}$ since $j = T\text{-pred}(i+1)$ and $\bar{\kappa} =$ crit $E^{\mathcal{T}}_i$. But then $i^{\mathcal{T}}_{j,i+1} \restriction \nu_k =$ identity, so $\psi \restriction \nu_k = (\psi \circ i^{\mathcal{T}}_{j,i+1}) \restriction \nu_k = \pi_j \restriction \nu_k = \pi_k \restriction \nu_k$. This is why we do not need to re-define $\mathcal{E}^i \restriction j$.)

The existence of the cutoff point η_k of $\mathcal{R}_k$, for $j \leq k \leq i$, is not a problem because for each $k \in \omega$, only one such cutoff point is used. (It is used at stage $i+1$, where i is least such that $k \leq i$ and $i+1 \in [0,\omega]_T$.)

Let now, for $k \in \omega$

$$\mathcal{E}_k^\omega = \text{eventual value of} \quad \mathcal{E}_k^i \quad \text{as} \quad i \to \omega\,.$$

The eventual value exists since in fact $\mathcal{E}_k^i$ changes value at most once. Set

$$\begin{aligned} \mathcal{R}_\omega^\omega = \mathcal{R}_0^0 &= \text{common value of} \quad \mathcal{R}_j^i\,, \text{ for } \quad j \in [0,\omega]_T \quad \text{and} \quad i \geq j\,, \\ Q_\omega^\omega = Q_0^0 &= \text{common value of} \quad Q_j^i\,, \text{ for } j \in [0,\omega]_T\,, \\ \pi_\omega^\omega(i_{j,\omega}^{\mathcal{T}}(x)) &= \pi_j^\omega(x), \quad \text{for} \quad j \in [0,\omega]_T\,. \end{aligned}$$

(Here $\mathcal{E}_j^\omega = (\pi_j^\omega, Q_j^\omega, \mathcal{R}_j^\omega)$.) Then $\mathcal{E}^\omega = \langle(\pi_k^\omega, Q_k^\omega, \mathcal{R}_k^\omega) \mid k \leq \omega\rangle$ is the desired enlargement of $\mathcal{T}$.

The extension of this method of enlargement to arbitrary simple countable trees involves only more bookkeeping. The reader can see similar bookkeeping problems handled in [IT]. In one respect, in fact, what we are doing now is simpler than what is done in [IT]. In the present construction, all models at the enlargement level, i.e. all $\mathcal{R}_\beta^\alpha$'s, are ω-closed. This is true even for $\beta \geq \omega$. The construction of [IT] did not have this property, and that led to complications.

The techniques of §12 of [FSIT] allow one to drop our simplifying assumptions (2) and (3). This completes our sketch of the proof of Theorem 2.5. □

Putting together Lemma 2.4 and Theorem 2.5, we get

Corollary 2.7. *Suppose there is no proper class inner model with a Woodin cardinal; then for all $\eta \leq \Omega$ and $k \leq \omega$, $\mathfrak{C}_k(\mathcal{N}_\eta)$ exists and is k-iterable.*

Proof. The reader can find k-iterability defined in 5.1.4 of [FSIT]. Roughly speaking, it means: iterable with respect to simple, k-bounded iteration trees. The existence of $\mathfrak{C}_{k+1}(\mathcal{N}_\eta)$ comes from the k-iterability of $\mathfrak{C}_k(\mathcal{N}_\eta)$ via Theorem 8.1 of [FSIT]. (The Strong Uniqueness theorem, 6.2 of [FSIT], is used here to show that the iteration trees arising in the proof of 8.1 are simple.) So we need only show that if $\mathfrak{C}_k(\mathcal{N}_\eta)$ exists, then it is k-iterable.

Let $\mathcal{T}$ on $\mathfrak{C}_k(\mathcal{N}_\eta)$ be almost normal and k-maximal, and $\mathcal{T} \restriction \lambda$ simple for all $\lambda < lh\ \mathcal{T}$. Suppose $lh\ \mathcal{T}$ is a limit; the case $lh\ \mathcal{T}$ is a successor is similar. We want a cofinal wellfounded branch of $\mathcal{T}$, and from 2.4 and 2.5 we get such a branch in $V^{\mathrm{Col}(\omega,\kappa)}$ for all sufficiently large κ. So if $\mathcal{T}$ is simple, we are done. If not, then letting $\mathcal{M} = \mathfrak{C}_k(\mathcal{N}_\eta)$, we have δ such that $\mathcal{M} \models \delta$ is Woodin, and $\rho_{k+1}(\mathcal{M}) \geq \delta$. Let $\langle\gamma, e\rangle$ be lexicographically least above $\langle\eta, k\rangle$ such that $\mathfrak{C}_e(\mathcal{N}_\gamma) \models \delta$ is not Woodin, or $\rho_{e+1}(\mathcal{N}_\gamma) < \delta$. Such a pair $\langle\gamma, e\rangle$ must exist because otherwise δ is Woodin in a proper class inner model. $\mathcal{M}$ is an initial segment of $\mathfrak{C}_e(\mathcal{N}_\gamma)$, and all extenders from the $\mathcal{M}$-sequence have length $< \delta$, and δ is a cardinal of $\mathfrak{C}_e(\mathcal{N}_\gamma)$. So $\mathcal{T}$ lifts to a j-maximal iteration tree $\mathcal{T}^*$ on $\mathfrak{C}_e(\mathcal{N}_\gamma)$. Since $\mathcal{T}^*$ is simple, it has a cofinal wellfounded branch in V, and thus so does $\mathcal{T}$. □

Corollary 2.8. *If there is no proper class inner model with a Woodin cardinal, then $K^c = \mathcal{N}_\Omega$ exists.*

From 2.7 we get that if there is no proper class model with a Woodin cardinal, then player II wins the full iteration game on any $\mathfrak{C}_k(\mathcal{N}_\eta)$. His winning "iteration strategy" is just to pick the unique cofinal wellfounded branch (after perhaps extending $\mathfrak{C}_k(\mathcal{N}_\eta)$ to some larger $\mathfrak{C}_e(\mathcal{N}_\gamma)$.) The existence of iteration strategies is more important than their nature, and indeed once one gets to premice which are not 1-small, there may be more than one cofinal wellfounded branch from which to choose. In order to state the results of §3 - §5 in their proper generality, we make the following definitions. Since we require only normal, ω-maximal trees in §3 - §5, we restrict ourselves to these.

The full iteration game $\mathcal{G}(\mathcal{M},\theta)$ of §5 of [IT] has an obvious counterpart $\mathcal{G}^*(\mathcal{M},\theta)$ for "fine-structural" premice. In $\mathcal{G}^*(\mathcal{M},\theta)$, I and II build together a normal, ω-maximal tree $\mathcal{T}$ on $\mathcal{M}$. At move $\alpha+1<\theta$, I picks an extender $E_\alpha^\mathcal{T}$ on the $\mathcal{M}_\alpha^\mathcal{T}$ sequence such that $\gamma<\alpha \Rightarrow lh\ E_\gamma^\mathcal{T} < lh\ E_\alpha^\mathcal{T}$. The rules for ω-maximal trees then determine a $\beta\le\alpha$ such that $\beta = T\text{-pred}(\alpha+1)$, and an initial segment $\mathcal{M}_{\alpha+1}^*$ of $\mathcal{M}_\beta^\mathcal{T}$ and $k\le\omega$ such that $\mathcal{M}_{\alpha+1}^\mathcal{T} = \text{Ult}_k(\mathcal{M}_{\alpha+1}^*, E_\alpha^\mathcal{T})$. If $\mathcal{M}_{\alpha+1}^\mathcal{T}$ is illfounded, the game is over and I has won. At move $\lambda<\theta$, where λ is a limit, II must pick a cofinal branch b of $\mathcal{T}\restriction\lambda$ such that $D^\mathcal{T}\cap b$ is finite and $\mathcal{M}_b^\mathcal{T}$ is wellfounded; if he fails to do so, then I wins. If he succeeds in doing so, we set $\mathcal{M}_\lambda^\mathcal{T} = \mathcal{M}_b^\mathcal{T}$ and continue play. If I does not win $\mathcal{G}^*(\mathcal{M},\theta)$ at some move $\alpha<\theta$ for one of the reasons just given, then II wins.

We also want to consider a variant of this game which allows almost normal iteration trees to be played. Let $\mathcal{G}^*(\mathcal{M},(\omega,\theta))$ be played as follows. There are ω rounds. Round 1 is just a play of $\mathcal{G}^*(\mathcal{M},\theta)$, except that I must say "exit" at some move $\alpha+1<\theta$. (If he doesn't do so, and II doesn't lose $\mathcal{G}^*(\mathcal{M},\theta)$, then II has already won $\mathcal{G}^*(\mathcal{M},(\omega,\theta))$ after round 1.) If I says "exit" at $\alpha+1$, then play moves to round 2, and in this round I and II play $\mathcal{G}^*(\mathcal{M}_\alpha^{\mathcal{T}_1},\theta)$, where $\mathcal{T}_1$ is the tree produced in round 1. Once again, I must exit at some $\beta+1<\theta$, etc. If no one loses during the ω rounds, then we say that II has won $\mathcal{G}^*(\mathcal{M},(\omega,\theta))$.

Definition 2.9. *A premouse $\mathcal{M}$ is θ-iterable iff II has a winning strategy in $\mathcal{G}^*(\mathcal{M},\theta)$. A winning strategy for II in $\mathcal{G}^*(\mathcal{M},\theta)$ is a θ-iteration strategy for $\mathcal{M}$. Similarly, $\mathcal{M}$ is (ω,θ)-iterable iff II has a winning strategy in $\mathcal{G}^*(\mathcal{M},(\omega,\theta))$ and we call such a strategy an (ω,θ)-iteration strategy.*

An obvious copying construction gives

Lemma 2.10. *If $\mathcal{M}\preceq\mathcal{N}$ and $\mathcal{N}$ is θ-iterable, then $\mathcal{M}$ is θ-iterable. Similarly, if $\mathcal{M}\preceq\mathcal{N}$ and $\mathcal{N}$ is (ω,θ) iterable, then so is $\mathcal{M}$.*

From 2.7 we get

Theorem 2.11. *Suppose there is no proper class model with a Woodin cardinal; then K^c is (ω, θ)-iterable for all θ.*

We shall only make use of the $(\omega, \Omega + 1)$-iterability of K^c. We can prove this without assuming there are no proper class models with a Woodin cardinal, but using instead the measurability of Ω and assuming $K^c \models$ there are no Woodin cardinals. More precisely, we use that $A^\sharp$ exists for all sets $A \in V_\Omega$ in order to see that K^c is well-behaved with respect to trees $\mathcal{T} \in V_\Omega$ (using 2.4 (a)), and then the weak compactness of Ω to see that K^c is well behaved with respect to trees of length Ω. We use that $K^c \models$ there is no Woodin cardinal to show that the appropriate trees are simple, and thus have not just generic branches, but branches in V. In a similar vein, one can omit the hypothesis "there is no proper class model with a Woodin cardinal" in 2.8, by using the measurability of Ω. We have stated 2.8 and 2.11 as we have in order to point out what can be proved without using the measurability of Ω.

Most of the rest of this paper makes heavy use of Theorem 1.4, and we certainly do not know how to avoid the measurability of Ω as a hypothesis in that theorem. So we shall take "$K^c \models$ there is no Woodin cardinal" as our non-large-cardinal hypothesis, when we need one, instead of "there is no proper class model with a Woodin cardinal". We shall use:

Thorem 2.12. *Suppose $K^c \models$ there is no Woodin cardinal; then K^c is $(\omega, \Omega + 1)$- iterable.*

§3. Thick classes and universal weasels

We shall adapt Mitchell's notion of a thick class of ordinals to the present context.

Definition 3.1. *Let $\mathcal{M}$ and $\mathcal{N}$ be premice. A coiteration of $\mathcal{M}$ with $\mathcal{N}$ is a pair $(\mathcal{T},\mathcal{U})$ of normal, ω-maximal iteration trees such that $\mathcal{T}$ is on $\mathcal{M}$, $\mathcal{U}$ is on $\mathcal{N}$, and successor steps in the formation of $\mathcal{T}$ and $\mathcal{U}$ are determined by iterating the least disagreement and the rules for normal, ω-maximal trees. A coiteration $(\mathcal{T},\mathcal{U})$ is terminal iff there is no extension of $(\mathcal{T},\mathcal{U})$ to a properly longer coiteration. A terminal coiteration $(\mathcal{T},\mathcal{U})$ is successful iff $\mathcal{T}$ and $\mathcal{U}$ have last models $\mathcal{P}$ and $\mathcal{Q}$, and $\mathcal{P} \trianglelefteq \mathcal{Q}$ or $\mathcal{Q} \trianglelefteq \mathcal{P}$.*

We write $\mathcal{P} \trianglelefteq \mathcal{Q}$ iff $\mathcal{P}$ is an initial segment of $\mathcal{Q}$; that is, $\exists\alpha(\mathcal{P} = \mathcal{J}_\alpha^{\mathcal{Q}})$. We should note that, in contrast to the conventions of [IT] and [FSIT], the iteration trees in a coiteration are ordinary, "unpadded", trees, and so may have different lengths. The reader can find a somewhat more formal rendering of definition 3.1 in [PW], 1.9. Notice that 3.1 requires nothing concerning the choice of branches at limit steps. If Σ and Γ are θ-iteration strategies for $\mathcal{M}$ and $\mathcal{N}$ respectively, then for all $\alpha < \theta$ there is exactly one coiteration $(\mathcal{T},\mathcal{U})$ of $\mathcal{M}$ with $\mathcal{N}$ such that $\alpha = \max(lh\ \mathcal{T}, lh\ \mathcal{U})$ and $\mathcal{T}$ is a play by Σ and $\mathcal{U}$ is a play by Γ; moreover, if this coiteration is terminal then it is successful. We say $(\mathcal{T},\mathcal{U})$ is *determined by* (Σ,Γ), or a (Σ,Γ)*-coiteration* of $\mathcal{M}$ with $\mathcal{N}$.

We shall have little use for premice $\mathcal{M}$ such that $\Omega < \mathrm{OR} \cap \mathcal{M}$, so we make the following definition.

Definition 3.2. *A weasel is a premouse $\mathcal{M}$ such that $OR \cap \mathcal{M} = \Omega$. A set premouse is a premouse $\mathcal{M}$ such that $OR \cap \mathcal{M} < \Omega$. $\mathcal{M}$ is a proper premouse iff $\mathcal{M}$ is either a weasel or a set premouse.*

Any weasel is a model of ZFC, so no fine structure is used in forming ultrapowers of it.

The strong inaccessiblity of Ω gives us the following lemma, whose well known proof we omit.

Lemma 3.3. *Let $\mathcal{M}$ and $\mathcal{N}$ be proper premice, and let $(\mathcal{T},\mathcal{U})$ be a coiteration of $\mathcal{M}$ with $\mathcal{N}$ such that $lh\ \mathcal{T} = \theta+1$ and $lh\ U = \gamma+1$. Then $max(\theta,\gamma) \leq \Omega$, and if $max\ (\theta,\gamma) = \Omega$, then either*

(a) *$\mathcal{M}$ is a weasel, $D^{\mathcal{T}} \cap [0,\theta]_T = \emptyset$, $i_{0,\theta}^{\mathcal{T}}{}''\Omega \subseteq \Omega$, and $\mathcal{M}_\theta^{\mathcal{T}} \trianglelefteq \mathcal{N}_\gamma^{\mathcal{U}}$, or*

(b) *$\mathcal{N}$ is a weasel, $D^{\mathcal{U}} \cap [0,\gamma]_U = \emptyset$, $i_{0,\gamma}^{\mathcal{U}}{}''\Omega \subseteq \Omega$, and $\mathcal{N}_\gamma^{\mathcal{U}} \trianglelefteq \mathcal{M}_\theta^{\mathcal{T}}$.*

By Lemma 3.3, the coiteration determined by $\Omega+1$-iteration strategies for two $\Omega+1$-iterable proper premice must terminate successfully at some stage $\leq \Omega$. Moreover, at least one of the two premice iterates to a structure of height $\leq \Omega$.

Definition 3.4. *A weasel $\mathcal{M}$ is universal iff $\mathcal{M}$ is $\Omega+1$-iterable, and whenever $(\mathcal{T},\mathcal{U})$ is a coiteration of $\mathcal{M}$ with some proper premouse $\mathcal{N}$ such that $lh\ \mathcal{U} = \Omega+1$, then $\mathcal{N}$ is a weasel, $D^{\mathcal{U}} \cap [0,\Omega]_U = \emptyset$, $\forall\alpha < \Omega(i^{\mathcal{U}}_{0,\Omega}(\alpha) < \Omega)$, and $\mathcal{N}^{\mathcal{U}}_{\Omega} \trianglelefteq \mathcal{M}^{\mathcal{T}}_{lh\,\mathcal{T}-1}$.*

A weasel is universal just in case it is maximal in the mouse prewellorder $\leq^*$ on $\Omega+1$ iterable proper premice. (Roughly, $\mathcal{M} \leq^* \mathcal{N}$ iff $\mathcal{M}$ iterates to an initial segment of an iterate of $\mathcal{N}$. In order to give a precise definition and prove that $\leq^*$ is a prewellorder, one must impose a bound on the large cardinals in the proper premice being ordered. This is because one needs a simplicity hypothesis in the Dodd-Jensen lemma. Here we can assume that there is no proper class model with a Woodin cardinal.) Any two universal weasels are $\equiv^*$, that is, they have a common iterate. By 3.3, this common iterate is itself a universal weasel. If $\mathcal{M}$ is universal, then any $\Omega+1$ iterable set premouse $\mathcal{N}$ is strictly below $\mathcal{M}$ in the mouse order, and in fact a coiteration determined by $\Omega+1$ iteration strategies for $\mathcal{M}$ and $\mathcal{N}$ must terminate successfully at some stage $\alpha < \Omega$.

If there is no inner model with a strong cardinal, then any $\Omega+1$ iterable weasel which is strictly above all $\Omega+1$ iterable set premice in the mouse order is universal. (This fact was noticed by Jensen, who made it his definition of universality.) If our weasels can have strong cardinals, however, then this condition no longer suffices for universality. For suppose $\mathcal{P}$ is universal, and $\mathcal{P} \models \kappa$ is strong. Let Q come from hitting some image of each extender from $\mathcal{P}$ with critical point some image of κ once. Let $\mathcal{M} = \mathcal{J}^{Q}_{\Omega}$. Then every $\Omega+1$ iterable set premouse is $<^* \mathcal{M}$, but $\mathcal{M}$ is not universal.

We now show K^c is universal.

Lemma 3.5. *Let W be an $\Omega+1$ iterable weasel such that for stationary many regular cardinals $\alpha < \Omega$, $\alpha^{+^W} = \alpha^+$; then W is universal.*

Proof. Suppose that $\mathcal{N}$ is a proper premouse which is a counterexample to the universality of W. Let $\mathcal{T}$ on W and $\mathcal{U}$ on $\mathcal{N}$ result from the comparison process, and fix $\alpha \in [0,\Omega]_U$ such that $i^{\mathcal{U}}_{\alpha\Omega}$ is defined (that is, there is no further dropping along $[0,\Omega]_U$) and fix κ such that $i^{\mathcal{U}}_{\alpha,\Omega}(\kappa) = \Omega$. (We have $lh\ \mathcal{U} = \Omega+1$, because otherwise, $lh\ \mathcal{U} = \gamma+1 < \Omega$ for some γ, in which case $\mathrm{OR}^{\mathcal{N}_\gamma} \leq \Omega$, so that $\mathcal{N}_\gamma$ is an initial segment of the last model of $\mathcal{T}$, contrary to hypothesis.)

These are club many $\beta \in [\alpha,\Omega]_U$ such that $i^{U}_{\alpha,\beta}(\kappa) = \beta$. Moreover, by 3.3 we have that $i^{\mathcal{T}}_{0,\Omega}$ is defined, and $i^{\mathcal{T}}_{0,\Omega}(\Omega) = \Omega$. Thus there are club many $\beta \in [0,\Omega]_T$ such that $i^{\mathcal{T}''}_{0,\beta}\beta \subseteq \beta$.

Now let β be a regular cardinal in both the clubs of the last paragraph, and $\beta^{+^W} = \beta^+$. As β is regular and $i^{\mathcal{T}''}_{0\beta}\beta \subseteq \beta$, $i^{\mathcal{T}}_{0\beta}(\beta) = \beta$, and thus $i^{\mathcal{T}}_{0\beta}(\beta^+) = \beta^+ = (\beta^+)^{W_\beta}$. On the other hand, $(\beta^+)^{\mathcal{N}_\beta} = i^{\mathcal{U}}_{\alpha\beta}(\kappa^{+^{\mathcal{N}_\alpha}})$, and so $(\beta^+)^{\mathcal{N}_\beta} < \beta^+$ as $(\kappa^+)^{\mathcal{N}_\alpha} < \beta^+$. But then, as $\beta = \mathrm{crit}\ i^{\mathcal{U}}_{\beta\Omega}$ and $\beta \leq \mathrm{crit}\ i^{\mathcal{T}}_{\beta\Omega}$, $(\beta^+)^{\mathcal{N}_\Omega} =$

$(\beta^+)^{\mathcal{N}_\beta}$ and $(\beta^+)^{W_\Omega} = (\beta^+)^{W_\beta}$. It follows that $(\beta^+)^{\mathcal{N}_\Omega} < (\beta^+)^{W_\Omega}$, and this contradicts the fact that W_Ω is an initial segment of $\mathcal{N}_\Omega$. □

Remark. Below a strong cardinal, if $(\alpha^+)^W = \alpha^+$ for cofinally many $\alpha < \Omega$, then W is universal. This is no longer true past a strong cardinal, as a modification of the previous example shows.

Corollary 3.6. *If $K^c \models$ there is no Woodin cardinal, then K^c is universal.*

Proof. By Theorem 1.4, $(\alpha^+)^{K^c} = \alpha^+$ for μ- a.e. $\alpha < \Omega$, and by 2.12 K^c is $\Omega + 1$ iterable, so Lemma 3.5 does the job. □

We can also prove a converse to 3.5, under the assumption that $K^c \models$ there are no Woodin cardinals.

Theorem 3.7. (1) *Let R and W be weasels having a common iterate, i.e., a successful coiteration $(\mathcal{T}, \mathcal{U})$ such that $\mathcal{T}$ and $\mathcal{U}$ have the same last model. Then for all but nonstationary many regular $\alpha < \Omega$, $(\alpha^+)^R = \alpha^+$ iff $(\alpha^+)^W = \alpha^+$.*

(2) *Suppose $K^c \models$ there are no Woodin cardinals; then for any $\Omega + 1$ iterable weasel W, W is universal iff $(\alpha^+)^W = \alpha^+$ for stationary many regular $\alpha < \Omega$.*

Proof. (1) Assume $lh\ \mathcal{T} = lh\ \mathcal{U} = \Omega + 1$ for notational simplicity. Then for all but nonstationary many regular $\alpha < \Omega$, $\alpha \leq \min(\text{crit}(i^{\mathcal{T}}_{\alpha\Omega}), \text{crit}(i^{\mathcal{U}}_{\alpha\Omega}))$, and thus $(\alpha^+)^{\mathcal{M}^{\mathcal{U}}_\alpha} = (\alpha^+)^{\mathcal{M}^{\mathcal{T}}_\alpha} = (\alpha^+)^Q$, where $Q = \mathcal{M}^{\mathcal{T}}_\Omega = \mathcal{M}^{\mathcal{U}}_\Omega$. But also $i^{\mathcal{T}}_{0\alpha}(\alpha) = \alpha = i^{\mathcal{U}}_{0\alpha}(\alpha)$ for all but nonstationary many regular α. For such α, $(\alpha^+)^R = \alpha^+$ iff $(\alpha^+)^{\mathcal{M}^{\mathcal{T}}_\alpha} = \alpha^+$, and $(\alpha^+)^W = \alpha^+$ iff $(\alpha^+)^{\mathcal{M}^{\mathcal{U}}_\alpha} = \alpha^+$. This gives (1).

(2) One direction is 3.5, and does not require the smallness hypothesis on K^c. Conversely, if K^c has no Woodin cardinals, then $(\alpha^+)^{K^c} = \alpha^+$ for stationary many $\alpha < \Omega$, and if W is universal then it has a common iterate with K^c. Thus we can apply (1). □

For weasels small enough that linear iteration suffices for comparison (e.g. weasels no initial segment of which has a measurable limit of strong cardinals), the conclusion of (1) can be strengthened to: $(\alpha^+)^R = (\alpha^+)^W$ for all but nonstationary many regular $\alpha < \Omega$. Moreover, there is a converse to this strengthening, for such "very small" weasels: if R and W are $\Omega + 1$ iterable and $(\alpha^+)^R = (\alpha^+)^W$ for stationary (equivalently, all but nonstationary) many regular $\alpha < \Omega$, then R and W have a common iterate. (In fact, $R \leq^* W$ iff $(\alpha^+)^R \leq (\alpha^+)^W$ for stationary, or equivalently all but nonstationary, many regular $\alpha < \Omega$. The proof is based on the fact that if $\kappa < \alpha$, where α is inaccessible, and i comes from a linear iteration of length α of R, then $|i(\kappa^+)|^R < (\alpha^+)^R$. This is true even if $i \notin R$, since we can embed i into a "universal" linear iteration which is in R. We do not know whether this strengthening of (1), or its converse, hold for arbitrary 1-small weasels, even assuming K^c has no Woodin cardinals. We conjecture that if R and W are $\Omega+1$ iterable weasels, then $R \leq^* W$ iff for club many $\alpha < \Omega$, α regular $\Rightarrow (\alpha^+)^R \leq (\alpha^+)^W$.

The following definition adapts Mitchell's notion of a thick class to our context. The notion is useful in showing that certain hulls and iterations preserve universality.

Definition 3.8. *Let M be a weasel, and S, $\Gamma \subseteq M$. We say that Γ is S-thick in M iff*

(1) *$S \subseteq \Omega$ and S is stationary in Ω, and*
(2) *for all but nonstationary many $\alpha \in S$:*
   (a) *α is inaccessible, $(\alpha^+)^M = \alpha^+$, and α is not the critical point of a total-on-M extender from the M-sequence, and*
   (b) *$\Gamma \cap \alpha^+$ contains an α-club, and $\alpha \in \Gamma$.*

Notice that if Ω is S-thick in M, and M is $\Omega+1$ - iterable, then M is universal. Notice also that Ω is A_0-thick in K^c, provided that K^c satisfies "There are no Woodin cardinals" (cf.§1). We care most about A_0-thick sets in what follows.

Lemma 3.9. *Suppose Ω is S-thick in M. Then the class of sets which are S-thick in M is an Ω-complete filter.*

Lemma 3.10. *Let $\pi : H \to M$ be elementary, where H and M are weasels, and suppose ran π is S-thick in M. Then $\{\alpha \mid \pi(\alpha) = \alpha\}$ is S-thick in both H and M.*

Lemma 3.11. *Let $\mathcal{T}$ be an iteration tree on the weasel M, and Ω be S-thick in M. Let $\lambda \leq \Omega$, and suppose there is no dropping along $[0,\lambda]_\mathcal{T}$ (and $\lambda <$ lh $\mathcal{T}$), so that $i^\mathcal{T}_{0,\lambda}$ is defined. Suppose $i^{\mathcal{T}}_{0,\lambda}{}''\Omega \subseteq \Omega$; then $\{\alpha < \Omega \mid \alpha = i^\mathcal{T}_{0,\lambda}(\alpha)\}$ is S thick in both M and M_λ.*

Lemmas 3.9 through 3.11 are quite easy to prove. For 3.11: this is clear if $\lambda < \Omega$, so let $\lambda = \Omega$. Then $[0,\lambda]_\mathcal{T}$ is club in Ω, so for the typical $\alpha \in S$, $\alpha \in [0,\lambda]_\mathcal{T}$ and $i^\mathcal{T}_{0,\alpha}{}''\alpha \subseteq \alpha$. As α is inaccessible, $i^\mathcal{T}_{0\alpha}(\alpha) = \alpha$. As α is not the critical point of a total-on-M extender, α is not the critical point of a total-on-M_α extender, so α is not the critical point of a total-on-M_α extender from M_β, any $\beta \geq \alpha$. Thus $\alpha^+ <$ crit $i^\mathcal{T}_{\alpha,\Omega}$, and $\{\gamma \mid i^\mathcal{T}_{0,\Omega}(\gamma) = \gamma\}$ contains α and an α-club subset of α^+.

From Theorem 1.4 we get immediately

Theorem 3.12. *Suppose $K^c \models$ there are no Woodin cardinals; then Ω is A_0-thick in K^c.*

§4. The hull and definability properties

Definition 4.1. *Let $\mathcal{M}$ be a premouse and $X \subseteq \mathcal{M}$. Then*

$$a \in H^{\mathcal{M}}(X) \Leftrightarrow \text{for some } s \in X^{<\omega} \text{ and formula } \varphi, \; a = \text{unique } v \text{ such that } \mathcal{M} \models \varphi[v, s].$$

Notice here that $H^{\mathcal{M}}(X)$ in the *uncollapsed* hull of X inside $\mathcal{M}$.

Definition 4.2. *Suppose Ω is S-thick in $\mathcal{M}$, and let $\alpha < \Omega$. We say that $\mathcal{M}$ has the S-hull property at α iff whenever Γ is S-thick in $\mathcal{M}$*

$$P(\alpha)^{\mathcal{M}} \subseteq \textit{transitive collapse of} \;\; H^{\mathcal{M}}(\alpha \cup \Gamma).$$

In his work on the core model for sequences of measures, Mitchell makes heavy use of a lemma which states (translated into our context) that if Ω is S-thick in $\mathcal{M}$ then $\mathcal{M}$ has the S-hull property at all $\alpha < \Omega$. This will fail as soon as we get past sequences of measures, as the following example shows.

Example 4.3. Suppose Ω is S-thick in $\mathcal{M}$. Let E be an extender from the $\mathcal{M}$ sequence which is total on $\mathcal{M}$, and $\kappa = \text{crit } E$. Suppose E has a generator $> \kappa$, and let ξ be the least such. (So $\kappa^{+^{\mathcal{M}}} < \xi$, and $E \restriction \xi = \dot{E}^{\mathcal{M}}_{\xi}$.) Now let $\mathcal{N} = \text{Ult}_0(\mathcal{M}, E) = \text{Ult}_\omega(\mathcal{M}, E)$. Then Ω is S-thick in $\mathcal{N}$. We claim that $\mathcal{N}$ fails to have the hull property at ξ. For let $i : \mathcal{M} \to \mathcal{N}$ be the canonical embedding and $\Gamma = \text{ran } i$. Thus Γ is S-thick in $\mathcal{N}$. Moreover we can factor i as follows:

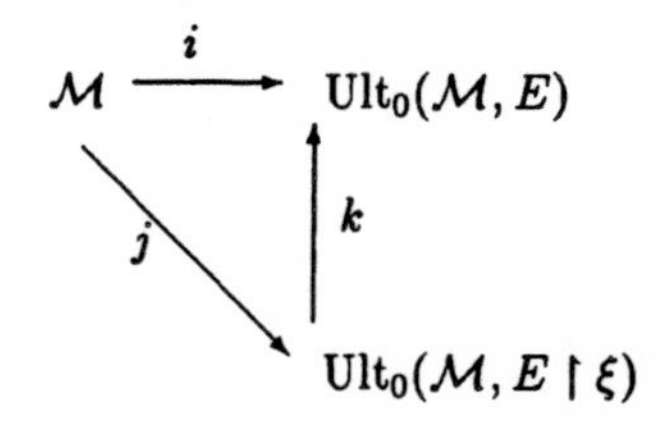

where $k([a, f]) = i(f)(a)$. We have $\xi = \text{crit } k$, and $\text{ran}(k) = H^{\mathcal{N}}(\xi \cup \Gamma)$, and so $Ult(\mathcal{M}, E \restriction \xi)$ is the transitive collapse of $H^{\mathcal{N}}(\xi \cup \Gamma)$. On the other hand, by coherence $\dot{E}^{\mathcal{N}}_{\xi} = \dot{E}^{\mathcal{M}}_{\xi} = E \restriction \xi$, so $E \restriction \xi \in \mathcal{N}$. As $E \restriction \xi$ is essentially a subset of ξ (in fact, of $(\kappa^+)^{\mathcal{N}}$) and $E \restriction \xi \notin \text{Ult}(\mathcal{M}, E \restriction \xi)$, we are done.

Remark. If Ω is S-thick in $\mathcal{M}$, and $\mathcal{P} = \mathcal{M}^{\mathcal{T}}_{\alpha}$ where $\mathcal{T}$ is an iteration tree on $\mathcal{M}$ and there is no dropping along $[0, \alpha]_T$ and $\alpha \leq \Omega$, and $\mathcal{M}$ has the S-hull property at all ξ, then $\mathcal{P}$ has the S-hull property at ξ iff for no $\beta + 1 \in [0, \alpha]_T$ do we have $(\kappa^+)^{\mathcal{M}_\beta} \leq \xi < \nu$, where $\nu = \nu(E^{\mathcal{T}}_{\beta})$ and $\kappa = \text{crit}(E^{\mathcal{T}}_{\beta})$. So we can recover from $\mathcal{P}$, using the hull property, the pairs (κ, ν) such that some extender with critical point κ and sup of generators $= \nu > (\kappa^+)^{\mathcal{P}}$ is used on the branch from $\mathcal{M}$ to $\mathcal{P}$. Notice also that $\mathcal{P}$ will have the S-hull property at club many $\xi < \Omega$.

Definition 4.4. *Let Ω be S-thick in $\mathcal{M}$, and $\alpha < \Omega$. We say $\mathcal{M}$ has the S-definability property at α iff whenever Γ is S-thick in $\mathcal{M}$, $\alpha \in H^{\mathcal{M}}(\alpha \cup \Gamma)$.*

Even at the sequences of measures level, it is possible that Ω is S-thick in $\mathcal{M}$, but $\mathcal{M}$ fails to have the S-definability property at some α. For let Ω be S-thick in $\mathcal{P}$, and $\mathcal{M} = \mathrm{Ult}(\mathcal{P},\mathcal{U})$ where $\mathcal{U}$ is total on $\mathcal{P}$ with critical point α.

In view of the previous examples, we cannot expect that K^c will have the A_0-hull or definability properties at all $\alpha < \Omega$. We shall show, however, that K^c has these properties at many $\alpha < \Omega$.

Lemma 4.5. *Let W be an $\Omega + 1$-iterable weasel, and let Ω be S-thick in W; then there is an elementary $\pi : M \to W$ such that ran π is S-thick in W, and M has the S-hull property at all $\alpha < \Omega$.*

Proof. Let us use "thick" to mean "S-thick", and "hull property" for "S-hull property". We shall define by induction on $\alpha \leq \Omega$ classes $N_\alpha \prec W$ such that N_α is thick in W. We shall have $N_{\alpha+1} \subseteq N_\alpha$ for all α, and $N_\lambda = \bigcap_{\beta<\lambda} N_\beta$ if λ is a limit. We then take ran π to be N_Ω.

In order to avoid dealing with collapse maps, let us say that a class $N \prec W$ which is thick in W has the hull property at κ, where $\kappa \in N$, iff $\bar{N}$ has the hull property at $\sigma(\kappa)$, where $\sigma : N \cong \bar{N}$ is the transitive collapse. Equivalently, N has the hull property at κ iff whenever $\Gamma \subseteq N$ is thick in W, and $A \subseteq \kappa$ and $A \in N$, then there is a set $B \in H^W((N \cap \kappa) \cup \Gamma)$ such that $B \cap \kappa = A$.

As we define the N_α's we define κ_α for $\alpha < \Omega$. κ_α will be the αth infinite cardinal of N_Ω. We shall have $N_\alpha \cap (\kappa_\alpha + 1) = N_\beta \cap (\kappa_\alpha + 1)$ for all $\beta > \alpha$. We also maintain inductively that N_β has the hull property at κ_α, for all $\beta > \alpha$.

Base step:

$$N_0 = W,$$
$$\kappa_0 = \omega.$$

Limit step:

$$N_\lambda = \bigcap_{\beta<\lambda} N_\beta$$
$$\kappa_\lambda = \text{least } \kappa \in N_\lambda \text{ such that } \kappa_\beta < \kappa \text{ for all } \beta < \lambda.$$

(By induction, κ_β is a cardinal of N_λ for all $\beta < \lambda$. So κ_λ is a cardinal of N_λ.)

Successor step: Suppose we are given N_α and κ_α, where $N_\alpha \models \kappa_\alpha$ is a cardinal. For each $A \subseteq \kappa_\alpha$ such that $A \in N_\alpha$ and A is a counterexample to N_α having the hull property at κ_α, pick a thick class Γ_A witnessing this. Let

$$\Gamma = \bigcap \{\Gamma_A \mid A \subseteq \kappa_\alpha \wedge A \in N_\alpha \wedge \Gamma_A \text{ exists}\},$$

where we set $\Gamma = N_\alpha$ if no Γ_A's exist, i.e. if N_α has the hull property at κ_α. Set

$$N_{\alpha+1} = H^W((N_\alpha \cap (\kappa_\alpha + 1)) \cup \Gamma).$$

(Each $\Gamma_A \subseteq N_\alpha$, so $N_{\alpha+1} \subseteq N_\alpha$.)

This finishes the construction. It is clear that if $\alpha < \beta < \Omega$, then $N_\alpha \cap (\kappa_\alpha+1) = N_\beta \cap (\kappa_\alpha+1)$, κ_α is a cardinal of N_β, and N_β has the hull property at κ_α. Moreover, $\langle \kappa_\gamma \mid \gamma \leq \beta \rangle$ is an initial segment of the cardinals of N_β. Moreover, N_β is thick.

Set $N_\Omega = \bigcap_{\alpha<\Omega} N_\alpha$. The assertions of the last paragraph are also obvious for $\beta = \Omega$, except that N_Ω is not obviously thick in W. The following claim is the key to showing this.

Claim. Let $\lambda < \Omega$ be a limit; then N_λ has the hull property at κ_λ, and therefore $N_{\lambda+1} = N_\lambda$.

Proof. Let M be the transitive collapse of N_λ, and κ the image of κ_λ under collapse. So κ is a limit cardinal of M, and M has the hull property at all $\alpha < \kappa$. We want to show that M has the hull property at κ. Notice Ω is thick in M.

Let Γ be thick in M, and $H =$ transitive collapse of $H^M(\kappa \cup \Gamma)$. We are to show $P(\kappa) \cap M \subseteq H$.

Let $\mathcal{T}$ on H and $\mathcal{U}$ on M be the iteration trees resulting from a coiteration of H with M determined by $\Omega + 1$ iteration strategies. (Notice that H and M are $\Omega + 1$-iterable because they are embeddable in W, and by 3.3 the comparison ends as a stage $\leq \Omega$.) Let $lh\ \mathcal{T} = \gamma + 1$ and $lh\ \mathcal{U} = \theta + 1$, where $\gamma, \theta \leq \Omega$ by 3.3.

Since H and M are both universal, $H_\gamma = M_\theta$ (where these are the final models or the two trees), and $i^{\mathcal{T}}_{0,\gamma}$ and $i^{\mathcal{U}}_{0,\theta}$ are both defined.

It is enough to see that crit $i^{\mathcal{U}}_{0,\theta} \geq \kappa$, as then $P(\kappa) \cap M = P(\kappa) \cap M_\theta = P(\kappa) \cap H_\gamma \subseteq H$. So suppose that crit $i^{\mathcal{U}}_{0,\theta} = \mu < \kappa$. Notice $(\mu^+)^M < \kappa$.

Let E be the first extender used along $[0,\theta]_U$; that is, $E = E^{\mathcal{U}}_\eta$ where $\eta + 1 \in [0,\theta]_U$ and $\mathcal{U}$-pred$(\eta+1) = 0$. So crit $E = \mu$ and $lh\ E \geq \kappa$. The argument of example 4.3 shows that $E \upharpoonright (\mu^+)^M$ witnesses that M_θ doesn't have the hull property at $(\mu^+)^M = (\mu^+)^{M_\theta}$. On the other hand, M and hence M_θ has the hull property at all ordinals $< (\mu^+)^M$.

If crit $i^{\mathcal{T}}_{0,\gamma} \geq (\mu^+)^M = (\mu^+)^H$, then $H_\theta = M_\theta$ has the hull property at $(\mu^+)^M$. Thus crit $i^{\mathcal{T}}_{0,\gamma} =$ crit $i^{\mathcal{U}}_{0,\theta} = \mu$.

Now let $A \subseteq \mu$ and $A \in M$. Let $\Gamma = \{\alpha \mid i^{\mathcal{T}}_{0,\gamma}(\alpha) = i^{\mathcal{U}}_{0,\theta}(\alpha) = \alpha\}$. By 3.9 and 3.11, Γ is thick (in H, M, and $H_\gamma = M_\theta$). So we can find a term τ such that

$$A = \tau^M[\bar{\beta}, \bar{c}] \cap \mu$$

where $\bar{\beta} \in \mu^{<\omega}$ and $\bar{c} \in \Gamma^{<\omega}$, using the hull property at μ in M. But then

$$i^{\mathcal{U}}_{0,\theta}(A) = \tau^{M_\theta}[\bar{\beta}, \bar{c}] \cap i^{\mathcal{U}}_{0,\theta}(\mu).$$

Now $\tau^H[\bar\beta,\bar c]\cap\mu=\tau^{H_\gamma}[\bar\beta,\bar c]\cap\mu=i^{\mathcal U}_{0,\theta}(A)\cap\mu=A$. Thus

$$i^{\mathcal T}_{0,\gamma}(A)=\tau^{H_\gamma}[\bar\beta,\bar c]\cap i^{\mathcal T}_{0,\gamma}(\mu)\,.$$

It follows that the 1st extenders used along $[0,\theta]_U$ and $[0,\gamma]_T$ agree up to the inf of the sups of their generators. This is a contradiction, as in the proof of the comparison lemma. This proves the claim. □

The claim implies N_Ω is thick in W. For by Födor's theorem, for all but nonstationary many $\alpha\in S$, $\alpha=\kappa_\alpha$ and for all $\beta<\alpha$ there is an α-club $C_\beta\subseteq N_\beta\cap\alpha^+$. Fix such an α. Let $C=\bigcap_{\beta<\alpha}C_\beta$; then C is α-club and $C\subseteq N_\alpha\cap\alpha^+$. But the claim tells us $N_\alpha=N_{\alpha+1}$, and hence $\kappa_{\alpha+1}=\alpha^+$. Since $N_{\alpha+1}\cap(\kappa_{\alpha+1}+1)=N_\Omega\cap(\kappa_{\alpha+1}+1)$, $C\subseteq N_\Omega\cap\alpha^+$.

This completes the proof of 4.5. □

We note in passing that the proof of the claim in 4.5 almost shows that if Ω is S-thick in M, then $\{\alpha<\Omega\mid M$ has the S-hull property at $\alpha\}$ is closed in Ω. It falls a bit short, however, and we do not know whether this is in fact true.

Lemma 4.6. *Let W be an $\Omega+1$ iterable weasel such that Ω is S-thick in W. Then for μ_0- a.e. $\alpha<\Omega$, W has the S-hull property at α.*

Proof. Let M be as given by lemma 4.5. Let $(\mathcal T,\mathcal U)$ be a coiteration of M with W determined by $\Omega+1$ iteration strategies. We suppose $lh\,\mathcal T=lh\,\mathcal U=\Omega+1$, the contrary case being very similar and left to the reader.

As both M and W are universal, there is no dropping on either $[0,\Omega]_T$ or $[0,\Omega]_U$, and $M_\Omega=W_\Omega$ (where these are the last models of $\mathcal T$ and $\mathcal U$ respectively). Let α be such that α is inaccessible, α is a limit point of $[0,\Omega]_T$ and $[0,\Omega]_U$, and $\forall\beta<\alpha$ ($lh\ E^{\mathcal T}_\beta<\alpha$ and $lh\ E^{\mathcal U}_\beta<\alpha$). Since branches of an iteration tree must be closed below their sup, all but nonstationary many inaccessible $\alpha<\Omega$ have these properties. Notice that $i^{\mathcal T}_{0,\alpha}(\alpha)=i^{\mathcal U}_{0,\alpha}(\alpha)=\alpha$, and that $\operatorname{crit}(i^{\mathcal T}_{\alpha,\Omega})\geq\alpha$ and $\operatorname{crit}(i^{\mathcal U}_{\alpha,\Omega})\geq\alpha$.

One can easily show that for any $\beta\in[0,\alpha]_T$, M_β has the hull property at η whenever $\sup\{lh\ E^{\mathcal T}_\gamma\mid\gamma+1\in[0,\beta]_T\}\leq\eta$. (Proof: let Γ be thick in M_β and let $A\subseteq\eta$, $A\in M_\beta$. Let $\eta^*\leq\eta$ be least such that $\eta\leq i^{\mathcal T}_{0\beta}(\eta^*)$. There is a function $f\in M$, $f:[\eta^*]^{<\omega}\times\eta^*\to\{0,1\}$, and an $a\in[\eta]^{<\omega}$, such that the characteristic function χ_A of A is given by: for $\xi<\eta$, $\chi_A(\xi)=i^{\mathcal T}_{0\beta}(f)(a,\xi)$. By the hull property in M we can find $\bar\xi\in\Gamma^{<\omega}$ such that $i^{\mathcal T}_{0\beta}(\bar\xi)=\bar\xi$, $b\in[\eta^*]^{<\omega}$, and a term τ such that $f=\tau^M[b,\bar\xi]\restriction([\eta^*]^{<\omega}\times\eta^*)$. So $i^{\mathcal T}_{0\beta}(f)=\tau^{M_\beta}[i^{\mathcal T}_{0\beta}(b),\bar\xi]\restriction([i^{\mathcal T}_{0\beta}(\eta^*)]^{<\omega}\times[i_{0\beta}(\eta^*)]$. So for $\gamma<\eta$, $\chi_A(\gamma)=\tau^{M_\beta}[i^{\mathcal T}_{0\beta}(b),\bar\xi](a,\gamma)$. Since $i^{\mathcal T}_{0\beta}(b)\in[\eta]^{<\omega}$ by the leastness of η^*, A is in the collapse of $H^{M_\beta}(\eta\cup\Gamma)$.)

Thus M_α has the hull property at α. Since $\alpha\leq\operatorname{crit}\ i^{\mathcal T}_{\alpha,\Omega}$, $M_\Omega=W_\Omega$ has the hull property at α. Since $\alpha\leq\operatorname{crit}\ i^{\mathcal U}_{\alpha,\Omega}$, W_α has the hull property at α.

Now let Γ be thick in W and $A \subseteq \alpha$, $A \in W$. We can find $\bar{\xi} \in \Gamma^{<\omega}$ and $b \in \alpha^{<\omega}$ such that $i^{\mathcal{U}}_{0\alpha}(\bar{\xi}) = \bar{\xi}$ and

$$i^{\mathcal{U}}_{0\alpha}(A) = \tau^{W_\alpha}[b, \bar{\xi}] \cap \alpha$$

for some term τ. Letting b be least which works for $\bar{\xi}$, $i^{\mathcal{U}}_{0\alpha}(A)$, and $\alpha = i^{\mathcal{U}}_{0\alpha}(\alpha)$, b is definable from elements of ran $i^{\mathcal{U}}_{0\alpha}$, so $b = i^{\mathcal{U}}_{0\alpha}(c)$ where $c \in \alpha^{<\omega}$. Then $A = \tau^W[c, \bar{\xi}] \cap \alpha$, as desired.

Thus W has the hull property at all but nonstationary many inaccessible $\alpha < \Omega$. □

Corollary 4.7. *Suppose $K^c \models$ there are no Woodin cardinals; then K^c has the A_0-hull property at μ_0- a.e. $\alpha < \Omega$.*

Proof. This is immediate from 2.12, 3.12, and 4.6. □

One should not expect that 4.6 will hold in full generality for the definability property. For suppose that for μ_0-a.e. $\alpha < \Omega$, α is measurable in K^c. Let W be the iterate of K^c obtained by using one total-on-K^c order zero measure from each measurable cardinal of K^c once. Then Ω is A_0-thick in W, and W is $\Omega + 1$ iterable, but W does not have the A_0-definability property at μ_0-a.e. α. Nevertheless, one can get a positive result in the case $W = K^c$, and this result will be important in the construction of "true K".

Lemma 4.8. *Suppose $K^c \models$ there are no Woodin cardinals; then for μ_0-a.e. $\alpha < \Omega$, K^c has the A_0-definability property at α.*

Proof. Assume the lemma fails, and for μ_0-a.e. α pick Γ_α thick in K^c such that

$$\alpha \notin H^{K^c}(\alpha \cup \Gamma_\alpha).$$

We can also arrange that $\alpha < \beta \Rightarrow \Gamma_\alpha \supseteq \Gamma_\beta$.

Let $V_1 = \mathrm{Ult}(V, \mu_0)$, and $j : V \to V_1$ be the canonical embedding. Let $V_2 = \mathrm{Ult}(V_1, j(\mu_0))$, and $j_1 : V_1 \to V_2$ the canonical embedding. Let $\Omega_1 = j(\Omega)$ and $\Omega_2 = j_1(\Omega_1)$. Let $K_1 = j(K^c)$ and $K_2 = j_1(K_1)$.

In V_2, we consider the map

$$\pi : H \cong H^{K_2}(\Omega \cup (\Gamma_\Omega)^{V_2}) \prec K_2$$

which inverts the collapse. Since $\Omega \notin H^{K_2}(\Omega \cup \Gamma_\Omega^{V_2})$, crit $\pi = \Omega$. Since K_2 is satisfied to have the hull property at Ω in V_2, $P(\Omega)^{K_2} \subseteq H$. Let E_π be the length $\pi(\Omega)$ extender derived from π. So $E_\pi \in V_2$, and measures all sets in $P(\Omega)^{K_2}$. Not every $E_\pi \restriction \nu$, $\nu \prec \pi(\Omega)$, belongs to K_2, as otherwise Ω is Shelah in K_2.

Claim. $E_\pi = E_j \cap ([\pi(\Omega)]^{<\omega} \times P(\Omega)^{K_2})$, where E_j is the extender derived from j.

Granted this claim, we can just repeat the proof of the main claim in the proof of Theorem 1.4 to get a contradiction. The point is that V_2 has suitable "background certificates" for the relevant fragments of E_j, so working in V_2 we get that every $E_\pi \restriction \nu$ is "on" the K_2 sequence in the right sense of "on". (1-smallness in no barrier to putting them on, as $K_2 \models$ there are no Woodins.)

Aside. Why isn't this an outright contradiction? We don't get $E_j \cap ([j(\Omega)]^{<\omega} \times P(\Omega)^{K_2})$ as member of V_2 without our false hypotheses.

Proof of Claim. Let $A \subseteq \Omega$ and $A \in K_2$. We must show that $\pi(A) = j(A) \cap \pi(\Omega)$. (It is easy to see that $\Gamma_\Omega^{V_2} \cap \Omega_1 \neq \emptyset$, and therefore $\pi(\Omega) < \Omega_1$.)

By the hull property for K_1 at Ω in V_1, we can find a term τ such that

$$A = \tau^{K_1}[\bar{c}, \bar{d}] \cap \Omega\,,$$

for some $\bar{c} \in \Omega^{<\omega}$ and $\bar{d} \in (\Gamma_\Omega^{V_1})^{<\omega}$. It follows that

$$j(A) = \tau^{K_2}[\bar{c}, j(\bar{d})] \cap \Omega_1\,.$$

Here we use that $j \circ j = j_1 \circ j$, so that $j(K_1) = K_2$. This also implies that $j(\Gamma^{V_1}) = \Gamma^{V_2}$, so that $j(\Gamma_\Omega^{V_1}) = \Gamma_{\Omega_1}^{V_2}$. Thus $j(\bar{d}) \in (\Gamma_{\Omega_1}^{V_2})^{<\omega}$.

On the other hand $\Gamma_{\Omega_1}^{V_2} \subseteq \Gamma_\Omega^{V_2}$, and $j(A) \cap \Omega = A$, so

$$A = \tau^{K_2}[\bar{c}, j(\bar{d})] \cap \Omega\,,$$

where $\bar{c} \in \Omega^{<\omega}$ and $j(\bar{d}) \in (\Gamma_\Omega^{V_2})^{<\omega}$. Moreover, from the definition of π,

$$\pi(A) = \tau^{K_2}[\bar{c}, j(\bar{d})] \cap \pi(\Omega)\,.$$

As $\pi(\Omega) < \Omega_1$, $\pi(A) = j(A) \cap \pi(\Omega)$, as desired. $\square$

§5. The construction of true K

The model K^c constructed in §1 depends too heavily on the universe within which it is constructed to serve our purposes. In this section we isolate a certain Skolem hull K of K^c, and prove that $K^V = K^{V[G]}$ whenever G is generic over V for a poset $\mathbb{P} \in V_\Omega$. The uniqueness result underlying this fact descends ultimately from Kunen's proof of the uniqueness of $L[\mu]$ ([Ku1]), and is based on the following lemma.

Lemma 5.1. *Let M and N be weasels which have the S-hull and S-definability properties at all $\beta < \alpha$. Let $(\mathcal{T},\mathcal{U})$ be a successful coiteration of M with N, let W be the common last model of $\mathcal{T}$ and $\mathcal{U}$, and let $i : M \to W$ and $j : N \to W$ be the iteration maps. Then $i \restriction \alpha = j \restriction \alpha =$ identity.*

Proof. Suppose not, and let $\kappa = \inf(\text{crit}(i),\ \text{crit}(j))$. Without loss of generality, let $\kappa = \text{crit}(i)$. We claim first that $\kappa = \text{crit}(j)$. For let

$$\Delta = \{\gamma < \Omega \mid i(\gamma) = j(\gamma) = \gamma\},$$

and recall that Δ is S-thick in M and N. Now $\kappa \notin H^W(\Delta)$, since otherwise κ is the range of i. On the other hand, N has the S-definability property at κ since $\kappa < \alpha$. Thus $\kappa \in H^N(\Delta)$, and if $\kappa < \text{crit}(j)$, then $\kappa \in H^W(\Delta)$. So $\kappa = \text{crit}(j)$.

We can now finish the proof as in 4.5. Let $A \subseteq \kappa$ and $A \in M$; we claim that $A \in N$ and $i(A) \cap \nu = j(A) \cap \nu$, where $\nu = \inf(i(\kappa), j(\kappa))$. For by the S-hull property of M at κ, we can find $\bar{\beta} \in \Delta^{<\omega}$ and a Skolem term τ such that $A = \tau^M(\bar{\beta}) \cap \kappa$. (Notice that $\kappa \subseteq \Delta$.) But then $i(A) = \tau^W(\bar{\beta}) \cap i(\kappa)$, so $A = \tau^W(\bar{\beta}) \cap \kappa = j(\tau^N(\bar{\beta})) \cap \kappa$. Since $\text{crit}(j) = \kappa$, this implies that $A = \tau^N(\bar{\beta}) \cap \kappa$, so that $A \in N$. Also $j(A) = \tau^W(\bar{\beta}) \cap j(\kappa)$, and therefore $i(A) \cap \nu = j(A) \cap \nu$ where $\nu = \inf(i(\kappa), j(\kappa))$.

A symmetric proof shows that if $A \subseteq \kappa$ and $A \in N$, then $A \in M$ and $i(A) \cap \nu = j(A) \cap \nu$. Let E and F be the first extenders used on the branches M-to-W and N-to-W of $\mathcal{T}$ and $\mathcal{U}$ respectively, and let $\theta = \inf(\nu(E),\ \nu(F))$, so that $\theta < \nu$. Then $i_E(A) \cap \theta = i(A) \cap \theta = j(A) \cap \theta = i_F(A) \cap \theta$ for A in $P(\kappa)^M$. It follows that $E \restriction \theta = F \restriction \theta$; on the other hand, since $(\mathcal{T},\mathcal{U})$ is a coiteration, no extender used in $\mathcal{T}$ is compatible with any extender used in $\mathcal{U}$. This contradiction completes the proof. □

Corollary 5.2. *Let M be an $\Omega+1$ iterable weasel which has the S-definability property at all $\beta < \alpha$; then M has the S-hull property at α.*

Proof. By induction we may suppose M has the S-hull property at all $\beta < \alpha$. Let $A \subseteq \alpha$, let Γ be S-thick in M, and let N be the transitive collapse of $H^M(\alpha \cup \Gamma)$. We must show that $A \in N$. Now N is $\Omega + 1$ iterable since it embeds in M, and Ω is S-thick in N. Also, N has the S hull and definability properties at all $\beta < \alpha$. Let $(\mathcal{T},\mathcal{U})$ be a successful coiteration of M with N, with iteration maps $i : M \to W$ and $j : N \to W$. By 5.1, $i \restriction \alpha = j \restriction \alpha =$

identity. Then $A = i(A) \cap \alpha$, so $A \in W$. Since $\mathrm{crit}(j) \geq \alpha$, $A \in N$, as desired. $\square$

Definition 5.3. *Let $\mathcal{M}$ be a set premouse, and let $S \subseteq \Omega$. We say that $\mathcal{M}$ is S-sound iff there is an $\Omega + 1$ iterable weasel W such that*
(1) $\mathcal{M} \trianglelefteq W$,
(2) *Ω is S-thick in W, and*
(3) *W has the S-definability property at all $\beta \in OR \cap \mathcal{M}$.*

Condition (3) of 5.3 is equivalent to: for every S-thick Γ, $\mathrm{OR} \cap \mathcal{M} \subseteq H^W(\Gamma)$. This is simply because if β is least such that $\beta \notin H^W(\Gamma)$, then $\beta \notin H^W(\beta \cup \Gamma)$. Also, by 5.2, condition (3) implies that W has the S-hull property at all $\beta \leq \mathrm{OR} \cap \mathcal{M}$.

Corollary 5.4. *Let $\mathcal{M}$ and $\mathcal{N}$ be S-sound; then either $\mathcal{M} \trianglelefteq \mathcal{N}$ or $\mathcal{N} \trianglelefteq \mathcal{M}$.*

Proof. Let W and R be weasels witnessing the S-soundness of $\mathcal{M}$ and $\mathcal{N}$ respectively. Let $i : W \to T$ and $j : R \to T$ be the iteration maps coming from a coiteration using $\Omega + 1$ iteration strategies. Then if $\alpha = \inf(\mathrm{OR}^{\mathcal{M}}, \mathrm{OR}^{\mathcal{N}})$, Lemma 5.1 implies $i \upharpoonright \alpha = j \upharpoonright \alpha =$ identity. This means that $\mathcal{M} \trianglelefteq \mathcal{N}$ or $\mathcal{N} \trianglelefteq \mathcal{M}$. $\square$

Let $S \subseteq \Omega$ be such that, for some $\Omega + 1$ iterable weasel W, Ω is S-thick in W. Clearly, there are many S-sound premice: $\mathcal{J}^W_\omega$ is an example, and $\mathcal{J}^W_\alpha$ for $\alpha = \omega_1^W$ is a slightly less trivial one. By 5.4 there is a proper premouse $\mathcal{R}$ such that the S-sound mice are precisely the proper initial segments of $\mathcal{R}$. We now give an alternative construction of $\mathcal{R}$, one which shows that it is embeddable in W.

Definition 5.5. *Suppose Ω is S-thick in W. Then we put*

$$x \in \mathit{Def}(W, S) \Leftrightarrow \forall \Gamma (\Gamma \text{ is } S\text{-thick in } W \Rightarrow x \in H^W(\Gamma)).$$

Clearly, $\mathrm{Def}(W, S) \prec W$. (More precisely, $\mathrm{Def}(W, S)$ is the universe of an elementary substructure of W. Recall here that the language of W includes a predicate $\dot{E}$ for its extender sequence. Thus a more careful statement would be that $(\mathrm{Def}(W, S), \in \upharpoonright \mathrm{Def}(W, S), \dot{E}^W \cap \mathrm{Def}(W, S))$ is an elementary submodel of W.)

We now show that, up to isomorphism, $\mathrm{Def}(W, S)$ is independent of W.

Lemma 5.6. *Let Ω be S-thick in W, an let $i : W \to Q$ be the iteration map coming from an iteration tree on W; then $i'' \mathit{Def}(W, S) = \mathit{Def}(Q, S)$.*

Proof. Let $\Delta = \{\gamma < \Omega \mid i(\gamma) = \gamma\}$, so that Δ is S-thick in both W and Q. Suppose first $x \in \mathrm{Def}(W, S)$. Let Γ be S-thick in Q; then $\Gamma \cap \Delta$ is S-thick in W, so $x = \tau^W(\bar{\beta})$ for some $\bar{\beta} \in (\Gamma \cap \Delta)^{<\omega}$ and term τ. But then $i(x) = \tau^Q(\bar{\beta})$, so $i(x) \in H^Q(\Gamma)$. As Γ was arbitrary, $i(x) \in \mathrm{Def}(Q, S)$.

Suppose next that $y \in \mathrm{Def}(Q, S)$. Since Δ is S-thick in Q, we can find $\bar{\beta} \in \Delta^{<\omega}$ so that $y = \tau^Q(\bar{\beta})$ for some term τ. Then $y = i(x)$, where $x = \tau^W(\bar{\beta})$.

Now let Γ be S-thick in W. Then $\Gamma \cap \Delta$ is S-thick in Q, and so $i(x) = \tau^Q(\bar{\alpha})$ for some term τ and $\bar{\alpha} \in (\Gamma \cap \Delta)^{<\omega}$. But then $x = \tau^W(\bar{\alpha})$, and since Γ was arbitrary, we have $x \in \mathrm{Def}(W, S)$. □

Corollary 5.7. *Let P and Q be $\Omega+1$ iterable weasels such that Ω is S-thick in each. Then $\mathrm{Def}(P, S) \cong \mathrm{Def}(Q, S)$.*

Proof. Once again, we are identifying $\mathrm{Def}(P, S)$ with the elementary submodel of P having universe $\mathrm{Def}(P, S)$. To prove 5.7, let $i : P \to W$ and $j : Q \to W$ be given by coiteration; then by 5.6 $\mathrm{Def}(P, S) \cong \mathrm{Def}(W, S) \cong \mathrm{Def}(Q, S)$. □

Definition 5.8. *Suppose there is an $\Omega + 1$ iterable weasel W such that Ω is S-thick in W; then $K(S)$ is the common transitive collapse of $\mathrm{Def}(W, S)$ for all such weasels W.*

If there is no $\Omega + 1$ iterable weasel W such that Ω is S-thick in W, then $K(S)$ is undefined.

Lemma 5.9. *Suppose $K(S)$ is defined; then for any set premouse $\mathcal{M}$, $\mathcal{M}$ is S-sound iff $\mathcal{M} \trianglelefteq K(S)$.*

Proof. Let $\mathcal{M}$ be S-sound, as witnessed by the weasel W. Then $\mathrm{OR}^{\mathcal{M}} \subseteq \mathrm{Def}(W, S)$, as one can see by an easy induction on $\beta \in \mathrm{OR}^{\mathcal{M}}$. Thus $\mathcal{M} \subseteq \mathrm{Def}(W, S)$, and since $\mathcal{M}$ is transitive, $\mathcal{M} \trianglelefteq K(S)$.

Conversely, let $\mathcal{M} \trianglelefteq K(S)$. Let R be an $\Omega+1$ iterable weasel such that Ω is S-thick in R, and let $\pi : K(S) \to R$ be elementary with $\mathrm{ran}\, \pi = \mathrm{Def}(R, S)$. Let $\theta = \sup \pi'' \,\mathrm{OR}^{\mathcal{M}}$, and for each $\alpha \in \theta - \mathrm{ran}\, \pi$, let

$$\Gamma_\alpha = \text{some } S\text{ -thick } \Gamma \text{ such that } \alpha \notin H^R(\Gamma)\,.$$

Then $\bigcap_{\alpha<\theta} \Gamma_\alpha$ is S-thick in W, so $\mathrm{Def}(R, S) \subseteq H^R(\bigcap_{\alpha<\theta} \Gamma_\alpha)$, while $H^R(\bigcap_{\alpha<\theta} \Gamma_\alpha) \cap \theta = \mathrm{Def}(R, S) \cap \theta$ by construction. Thus if we set

$$W = \text{transitive collapse of } H^R\left(\bigcap_{\alpha<\theta} \Gamma_\alpha\right)$$

then W is an $\Omega + 1$ iterable weasel with Ω S-thick in W, and $\mathcal{M} \trianglelefteq W$. It is easy to see that W has the S-definability property at all $\beta \in \mathrm{OR}^{\mathcal{M}}$: if not, then letting $\sigma : W \to R$ invert the collapse, we have that R fails to have the S definability property at $\sigma(\beta)$. Since $\beta \in \mathrm{OR}^{\mathcal{M}}$, $\sigma(\beta) = \pi(\beta)$, and since $\pi(\beta) \in \mathrm{Def}(R, S)$, this is a contradiction. Thus W witnesses that $\mathcal{M}$ is S-sound. □

As far as we know, it could happen that $K(S)$ is defined (that is, there is an $\Omega + 1$ iterable weasel W such that Ω is S-thick in W) and yet $K(S)$ is a set premouse, and hence not universal. We now show that if K^c satisfies "there are no Woodin cardinals", then $K(A_0)$, which exists by 2.12 and 3.12, is a universal weasel.

Theorem 5.10. *Suppose that $K^c \models$ there are no Woodin cardinals; then $K(A_0)$ is a weasel, and moreover $(\alpha^+)^{K(A_0)} = \alpha^+$ for μ_0–a.e. $\alpha < \Omega$, so that $K(A_0)$ is universal.*

Proof. We first show that $K(A_0)$ is a weasel, or equivalently, that $\text{Def}(K^c, A_0)$ is unbounded in Ω. So suppose otherwise toward a contradiction. It is easy then to see that there are A_0-thick classes Γ_ξ, for $\xi < \Omega$, such that

$$\xi < \delta \Rightarrow \Gamma_\delta \subseteq \Gamma_\xi ,$$

and letting

$$b_\xi = \text{least ordinal } \nu \in (H^{K^c}(\Gamma_\xi) - \text{Def}(K^c, A_0)) ,$$

we have that

$$(\text{Def}(K^c, A_0) \cup \Omega) \subseteq b_0 \text{ and } \xi < \delta \Rightarrow b_\xi < b_\delta .$$

By Lemma 4.8, we can fix ν such that $0 < \nu < \Omega$, $\nu = \sup\{b_\xi \mid \xi < \nu\}$, and K^c has the A_0-definability property at ν. Let $c \in \nu^{<\omega}$ and $d \in \Gamma_{\nu+1}$ and τ a term be such that

$$\nu = \tau^{K^c}[c, d] .$$

Fix $\xi < \nu$ such that $c \in b_\xi^{<\omega}$, so that

$$\exists c \in b_\xi^{<\omega}(b_\xi < \tau^{K^c}[c, d] < b_{\nu+1}) .$$

This is an assertion about b_ξ, d, and $b_{\nu+1}$, all of which belong to $H^{K^c}(\Gamma_\xi)$. Thus we can find $c^* \in (b_\xi \cap H^{K^c}(\Gamma_\xi))^{<\omega}$ such that

$$b_\xi < \tau^{K^c}[c^*, d] < b_{\nu+1} .$$

But $b_\xi \cap H^{K^c}(\Gamma_\xi) = \text{Def}(K^c, A_0) \cap \Omega$, so $c^* \in \text{Def}(K^c, A_0)$. This implies $\tau^{K^c}[c^*, d] \in H^{K^c}(\Gamma_{\nu+1})$, and since $\text{Def}(K^c, A_0) \subseteq b_0$, and $b_0 < \tau^{K^c}[c^*, d] < b_{\nu+1}$, this contradicts the definition of $b_{\nu+1}$.

Thus $\text{Def}(K^c, A_0)$ is unbounded in Ω. We claim that, in fact, $\text{Def}(K^c, A_0) \cap \Omega$ has μ_0- measure one. For this it is enough to show that if $\nu < \Omega$ is regular, $\text{Def}(K^c, A_0)$ is unbounded in ν, and K^c has the A_0-definability property at ν, then $\nu \in \text{Def}(K^c, A_0)$. So suppose ν is a counterexample to the last sentence.

For each $\eta \in (\nu + 1) - \text{Def}(K^c, A_0)$, pick an A_0-thick class Γ_η such that $\eta \notin H^{K^c}(\Gamma_\eta)$, and let $\Gamma = \bigcap_\eta \Gamma_\eta$. Let b be the least ordinal in $H^{K^c}(\Gamma)$ which is strictly greater than ν. Fix $\xi \in \text{Def}(K^c, A_0) \cap \nu$ and $d \in \Gamma^{<\omega}$ such that for some $c \in \xi^{<\omega}$ and term τ, $\nu = \tau^{K^c}[c, d]$. Then, as in the proof that $\text{Def}(K^c, A_0)$ is unbounded, for each $\eta \in \text{Def}(K^c, A_0) \cap \nu$ we can find $c_\eta \in \xi^{<\omega} \cap \text{Def}(K^c, A_0)$ such that $\eta < \tau^{K^c}[c_\eta, d] < b$. As ν is regular, we can fix c^* so that $c_\eta = c^*$ for arbitrarily large $\eta < \nu$. But then $\nu \leq \tau^{K^c}[c^*, d] < b$. Since $c^* \in \text{Def}(K^c, A_0) \subseteq H^{K^c}(\Gamma)$, this contradicts the definition of b.

Finally, we show that for μ_0-a.e. ν, Def(K^c, A_0) is unbounded in ν^+. This clearly implies that $(\nu^+)^{K(A_0)} = \nu^+$ for μ_0-a.e. ν, and so completes the proof of 5.10. So suppose not; then we can fix $\nu \in$ Def(K^c, A_0) such that $(v^+)^{K^c} = \nu^+$, K^c has the A_0-hull property at ν, and Def$(K^c, A_0) \cap \nu^+$ is bounded in ν^+. We have then an A_0-thick class Γ such that $H^{K^c}(\Gamma)$ is bounded in ν^+, say by $\delta < \nu^+$.

By the hull property we have a term τ and $d \in \Gamma^{<\omega}$ such that for some $c \in (\nu+1)^{<\omega}$

$$\delta < \tau^{K^c}[c, d] < \nu^+ .$$

But now, set

$$\eta = \sup\{\tau^{K^c}[c^*, d] \mid c^* \in (\nu+1)^{<\omega} \wedge \tau^{K^c}[c^*, d] < v^+\} .$$

Then $\delta < \eta < \nu^+$, and $\eta \in H^{K^c}(\Gamma)$ since $\nu, d \in H^{K^c}(\Gamma)$. This contradicts the choice of δ. □

It is very easy to show that, modulo the absoluteness of $\Omega + 1$ iterability, $K(S)$ is absolute under "set" forcing.

Theorem 5.11. *Suppose $K(S)$ is defined, as witnessed by the $\Omega+1$-iterable weasel W such that Ω is S-thick in W. Let G be V-generic over $\mathbb{P}$, where $\mathbb{P} \in V_\Omega$, and suppose that $V[G] \models W$ is $\Omega+1$-iterable. Then $V[G] \models$ "$K(S)$ exists, as witnessed by W", and $K(S)^{V[G]} = K(S)^V$.*

Proof. V and $V[G]$ have the same cardinals and cofinalities $> |\mathbb{P}|$; moreover, if $C \in V[G]$ and C is club in some regular $\nu > |\mathbb{P}|$, then $\exists D \in V$ ($D \subseteq C$ and D is club in ν). It follows that for any class $\Gamma \subseteq \Omega$ in $V[G]$

$$V[G] \models \Gamma \text{ is } S \text{ thick in } W \text{ iff } \exists \Delta \subseteq \Gamma (V \models \Delta \text{ is } S\text{-thick in } W) .$$

This implies that Ω is S-thick in W in $V[G]$, and that Def$(W, S)^{V[G]} =$ Def$(W, S)^V$. Since W is $\Omega+1$ iterable in $V[G]$ by hypothesis, we get that $K(S)^{V[G]}$ exists and $K(S)^{V[G]} = K(S)^V$. □

We doubt that one can show that $\Omega+1$-iterability of W is absolute for "set" forcing in the abstract, although we have no counterexample here. It seems likely that one must appeal to the existence of a definable $\Omega+1$ iteration strategy for W. This will come from a simplicity restriction on the iteration trees on W, which in turn will come from a smallness condition on W. At the one Woodin cardinal level, we can use the following lemma, whose proof is a slight extension of that of 2.4(a).

Lemma 5.12. *Let W be an $\Omega+1$-iterable (respectively, $(\omega, \Omega+1)$-iterable) proper premouse such that $W \models$ there are no Woodin cardinals, and let G be V-generic over $\mathbb{P}$, where $\mathbb{P} \in V_\Omega$. Then $V[G] \models W$ is $\Omega+1$ iterable (respectively, $(\omega, \Omega+1)$-iterable).*

Proof. We give the proof for $\Omega+1$-iterability. Using the weak compactness of Ω in $V[G]$, it is enough to show that $V[G]$ satisfies: whenever $\mathcal{T}$ is a putative normal, ω-maximal iteration tree on $\mathcal{J}_\alpha^W$, for some W-cardinal $\alpha < \Omega$, and $lh\ \mathcal{T} < \Omega$, then either $\mathcal{T}$ has a last, wellfounded model, or $\mathcal{T}$ has a cofinal wellfounded branch. So suppose $\mathcal{T}$ is a tree on $\mathcal{J}_\alpha^W$ which is a counterexample to this assertion, and let $\mathcal{T}, \mathcal{J}_\alpha^W \in V_\eta[G]$, where $\eta < \Omega$ is an inaccessible cardinal, and $\mathbb{P} \in V_\eta$. By the Löwenheim-Skolem theorem, we have in V a countable transitive M and elementary $\pi : M \to V_\eta$ such that $\mathcal{J}_\alpha^W, \mathbb{P} \in \text{ran}\ \pi$. Let $\pi(\langle \bar{W}, \bar{\mathbb{P}}\rangle) = \langle \mathcal{J}_\alpha^W, \mathbb{P}\rangle$; then M thinks that $\bar{\mathbb{P}}$ has a condition forcing the existence of a "bad" tree on $\bar{W}$. Since M is countable, we can find in V on M-generic filter $\bar{G}$ on $\bar{\mathbb{P}}$ such that $M[\bar{G}] \models \bar{\mathcal{T}}$ is a "bad" tree on $\bar{W}$. Notice that since $\bar{W}$ satisfies "There are no Woodin cardinals", $\bar{\mathcal{T}}$ is simple; moreover, since $\pi : \bar{W} \to \mathcal{J}_\alpha^W$ is elementary, $\bar{\mathcal{T}}$ is "good" in V. Thus $\bar{\mathcal{T}}$ cannot have a last, illfounded model, and $\bar{\mathcal{T}}$ has a unique cofinal wellfounded branch b in V. It is enough for a contradiction to show that $b \in M[\bar{G}]$, and for this it is enough to show $b \in M[\bar{G}][H]$, where H is $M[\bar{G}]$ generic for $\text{Col}(\omega, \max(|\bar{\mathcal{T}}|, |\bar{W}|)^{M[\bar{G}]})$. But now in $M[\bar{G}][H]$ there is a real x which codes $(\bar{\mathcal{T}}, \bar{W})$. Also, $x^\sharp \in M[\bar{G}][H]$, since M is closed under the sharp function on arbitrary sets because it embeds elementarily in V_η. It is a Σ_2^1 assertion about x that $\bar{\mathcal{T}}$ has a cofinal wellfounded branch, this assertion is true in V, and $x^\sharp \in M[\bar{G}][H]$, so this assertion is true in $M[\bar{G}][H]$. As b is unique, this means that $b \in M[\bar{G}][H]$. □

Putting together 5.11 and 5.12, we get

Theorem 5.13. *Suppose $K(S)$ is defined, as witnessed by a weasel W such that $W \models$ there are no Woodin cardinals. Let G be V-generic for $\mathbb{P}$, where $\mathbb{P} \in V_\Omega$. Then $V[G] \models$ "$K(S)$ is defined, as witnessed by W", and $K(S)^{V[G]} = K(S)^V$.*

Corollary 5.14. *Suppose $K^c \models$ there are no Woodin cardinals, and let G be V-generic over $\mathbb{P} \in V_\Omega$. Then $V[G] \models$ "$K(A_0)$ is defined, as witnessed by $(K^c)^V$; moreover $(\alpha^+)^{K(A_0)} = \alpha^+$ for μ_0- a.e. $\alpha < \Omega$".*

Let us observe in passing that if there is an $\Omega+1$ iterable weasel W such that Ω is S-thick in W, for some S, and $W \models$ there are no Woodin cardinals, then in fact $K^c \models$ there are no Woodin cardinals. [Sketch: If $K^c \models$ there is a Woodin cardinal, then its coherent sequence is of size $< \Omega$. Let $(\mathcal{T}, \mathcal{U})$ be a terminal coiteration of K^c with W, using an iteration strategy on the W side and picking unique cofinal branches on the K^c side. $(\mathcal{T}, \mathcal{U})$ cannot be successful, since otherwise the K^c side would have iterated past W, contrary to $(\alpha^+)^W = \alpha^+$ for stationary many α. Thus it must be that $\mathcal{T}$ has no cofinal wellfounded branch. The existence of generic branches for trees on K^c then implies $\delta(\mathcal{T})$, the sup of the lengths of the extenders used in $\mathcal{T}$, is Woodin in an iterate of W, a contradiction.] Thus we can add to the conclusion of 5.14: $(K^c)^{V[G]} \models$ there are no Woodin cardinals. We are not sure whether

Ω is $(A_0)^V$-thick in $(K^c)^{V[G]}$, however. We now show that, if there is an $(\omega, \Omega+1)$-iterable weasel, then there is at most one weasel of the form $K(S)$. First, let us note:

Lemma 5.15. *If there is an $(\omega, \Omega+1)$-iterable universal weasel, then every $\Omega+1$-iterable proper premouse is $(\omega, \Omega+1)$-iterable.*

Proof. Let W be universal and Σ an $(\omega, \Omega+1)$-iteration strategy for W. Let $\mathcal{M}$ be an $\Omega+1$ iterable premouse. By coiteration, we obtain a normal iteration tree $\mathcal{T}$ on W which is a play of round 1 of $\mathcal{G}^*(W, (\omega, \Omega+1))$ according to Σ, with last model $\mathcal{P}$, and an elementary $\pi : \mathcal{M} \to \mathcal{P}$. But then $\mathcal{P}$ is $(\omega, \Omega+1)$-iterable, and so by 2.9, so is $\mathcal{M}$. □

The next lemma says that, except possibly for its ordinal height, $K(S)$ is independent of S.

Lemma 5.16. *Suppose there is an $(\omega, \Omega+1)$-iterable universal weasel, and that S and T are stationary sets such that $K(S)$ and $K(T)$ exist. Then $K(S) \trianglelefteq K(T)$ or $K(T) \trianglelefteq K(S)$. In particular, if $K(S)$ and $K(T)$ are weasels, then $K(S) = K(T)$.*

Proof. Let $\mathcal{M}$ be S-sound, as witnessed by W, and let $\mathcal{N}$ be T-sound, as witnessed by R. We assume without loss of generality that $\mathrm{OR}^{\mathcal{M}} \leq \mathrm{OR}^{\mathcal{N}}$. W and R are $(\omega, \Omega+1)$-iterable by Lemma 5.15.

By Theorem 3.7 (1), for all but non-stationary many $\alpha \in S \cup T$, $(\alpha^+)^R = (\alpha^+)^W = \alpha^+$. Now let W^* be the (linear) iterate of W obtained by taking an ultrapower by the order zero total measure on α from W, for each $\alpha \in T - \mathrm{OR}^{\mathcal{M}}$ such that $W \models \alpha$ is measurable. Similarly, let R^* be obtained from R by taking an ultrapower by the order zero measure on α at each $\alpha \in S - \mathrm{OR}^{\mathcal{N}}$ such that $R \models \alpha$ is measurable. Then W^* and R^* still witness the S and T soundness of $\mathcal{M}$ and $\mathcal{N}$, respectively. Moreover, Ω is $S \cup T$ thick in each of W^* and R^*.

Let $i : W^* \to Q$ and $j : R^* \to Q$ come from coiteration. Let $\kappa = \min(\mathrm{crit}(i), \mathrm{crit}(j))$. It is enough to show that $\mathrm{OR}^{\mathcal{M}} \leq \kappa$, for then $\mathcal{M} \trianglelefteq \mathcal{N}$ as desired, so assume that $\kappa < \mathrm{OR}^{\mathcal{M}}$.

Suppose that $\kappa = \mathrm{crit}(i) < \mathrm{crit}(j)$. Since Ω is T-thick in R^* and W^*, and $\kappa \in \mathrm{Def}(R^*, T)$, we can find a term τ and common fixed points $\alpha_1 \cdots \alpha_k$ of i and j so that $\kappa = \tau^{R^*}[\bar{\alpha}]$. But then $\kappa = j(\kappa) = \tau^Q[\bar{\alpha}] = i(\tau^{W^*}[\bar{\alpha}])$, so $\kappa \in \mathrm{ran}(i)$, a contradiction. Similarly, we get $\mathrm{crit}(i) \leq \mathrm{crit}(j)$, so $\mathrm{crit}(j) = \mathrm{crit}(i) = \kappa$.

A similar argument with the hull property gives the usual contradiction. let $A \subseteq \kappa$ and $A \in W^*$. We have a term τ and common fixed points $\bar{\alpha}$ of i and j such that $A = \tau^{W^*}[\bar{\alpha}] \cap \kappa$, using here that W^* has the S-hull property as κ and Ω is S-thick in R^*. Then $i(A) = \tau^Q[\bar{\alpha}] \cap i(\kappa)$, so $\tau^Q[\bar{\alpha}] \cap \kappa = \tau^{R^*}[\bar{\alpha}] \cap \kappa = A$, and $j(A) = \tau^Q[\bar{\alpha}] \cap j(\kappa)$. Thus $i(A)$ and $j(A)$ agree below $\min(i(\kappa), j(\kappa))$. This implies that the extenders used first on the branches of the two trees in our coiteration which produced i and j are compatible with one another. This is a contradiction. □

Definition 5.17. *Suppose there is an $(\omega, \Omega+1)$ iterable universal weasel, and that $K(S)$ exists for some S; then we say that K exists, and define K to be the unique proper premouse $\mathcal{M}$ such that $\forall \mathcal{P}, S$ ($\mathcal{P}$ is S-sound $\Leftrightarrow \mathcal{P} \trianglelefteq \mathcal{M}$).*

We do know whether it is consistent with the definitions we have given that K exists, but is only a set premouse or a non-universal weasel. If we assume that $K^c \models$ there are no Woodin cardinals, then K exists by 2.12, 3.6, and 3.12; moreover K is universal by 5.10. We summarize what we have proved about K under this "no Woodin cardinals" assumption:

Theorem 5.18. *Suppose $K^c \models$ there are no Woodin cardinals; then*

(1) *K exists, and is $(\omega, \Omega+1)$ iterable,*

(2) *$(\alpha^+)^K = \alpha^+$ for μ_0- a.e. $\alpha < \Omega$, and*

(3) *if G is V-generic/$\mathbb{P}$, for some $\mathbb{P} \in V_\Omega$, then $V[G] \models$ "K exists, is $(\omega, \Omega+1)$ iterable, and $(\alpha^+)^K = \alpha^+$ for μ_0- a.e. $\alpha < \Omega$"; moreover $K^{V[G]} = K^V$.*

§6. An inductive definition of K

The definition of K given in 5.17 is $\Sigma_\omega(V_{\Omega+1})$, and therefore much too complicated for some purposes. In this section we shall give an inductive definition of K whose logical form is as simple as possible. Assuming that K^c has no Woodin cardinals, we shall show that $K \cap HC$ is $\Sigma_1(L_{\omega_1}(\mathbb{R}))$ in the codes; Woodin has shown that in general no simpler definition is possible.

The following notion is central to our inductive definition of K.

Definition 6.1. *Let $\mathcal{M}$ be a proper premouse such that $\mathcal{M} \models ZF - \{Powerset\}$ and $\mathcal{J}_\alpha^{\mathcal{M}}$ is S-sound. We say $\mathcal{M}$ is (α, S)-strong iff there is an $(\omega, \Omega+1)$ iterable weasel which witnesses that $\mathcal{J}_\alpha^{\mathcal{M}}$ is S-sound, and whenever W is a weasel which witnesses that $\mathcal{J}_\alpha^{\mathcal{M}}$ is S-sound, and Σ is an $(\omega, \Omega+1)$ iteration strategy for W, then there is a length $\theta+1$ iteration tree $\mathcal{T}$ on W which is a play by Σ and such that $\forall\gamma < \theta(\nu(E_\gamma^{\mathcal{T}}) \geq \alpha)$, and a $Q \trianglelefteq W_\theta^{\mathcal{T}}$, and a fully elementary $\pi : \mathcal{M} \to Q$ such that $\pi \restriction \alpha =$ identity.*

We shall see that it is possible to define "(α, S)-strong" by induction on α. First, let us notice:

Lemma 6.2. *Let W be an $(\omega, \Omega+1)$ iterable weasel which witnesses that $\mathcal{J}_\alpha^W$ is S-sound; then W is (α, S) strong.*

Proof. Let R be a weasel which witnesses $\mathcal{J}_\alpha^W$ is S-sound, and let Σ be an $\Omega+1$ iteration strategy for R. Let Γ be an $\Omega+1$ iteration strategy for W, and let $(\mathcal{T}, \mathcal{U})$ be the successful coiteration of R with W determined by (Σ, Γ). Let Q be the common last model of $\mathcal{T}$ and $\mathcal{U}$, and let $\pi : W \to Q$ be the iteration map given by $\mathcal{U}$. By Lemma 5.1, $\pi \restriction \alpha =$ identity. □

Lemma 6.2 admits the following slight improvement. Let W witness that $\mathcal{J}_\alpha^W$ is S-sound, and let Σ be an $(\omega, \Omega+1)$ iteration strategy for W. Let $\mathcal{T}$ be an iteration tree played by Σ such that $\forall\gamma < \theta(\nu(E_\gamma^{\mathcal{T}}) \geq \alpha)$, where $\theta+1 = lh\, \mathcal{T}$; then $W_\theta^{\mathcal{T}}$ is (α, S) strong. [Proof: Let R be any weasel witnessing $\mathcal{J}_\alpha^W$ is S-sound. Comparing R with W, we get an iteration tree $\mathcal{U}$ on R and a map $\pi : W \to R_\eta^{\mathcal{U}}$, where $\eta = lh\, \mathcal{U} - 1$. By 5.1, $\text{crit}(\pi) \geq \alpha$. Let $\sigma : W_\theta^{\mathcal{T}} \to (R_\eta^{\mathcal{U}})_\theta^{\pi\mathcal{T}}$ be the copy map. Then σ and $\mathcal{U}^\frown \pi\mathcal{T}$ are as required in 6.1 for R.] This shows that we obtain a definition of (α, S) strength equivalent to 6.1 if we replace "whenever W is a weasel" by "there is a weasel W" in 6.1. It also shows that there are (α, S) strong weasels other than those described in 6.2. For example, suppose W witnesses that $\mathcal{J}_\alpha^W$ is S-sound, and E is an extender on the W sequence which is total on W and such that $\text{crit}(E) < \alpha \leq \nu(E)$. Setting $R = \text{Ult}(W, E)$, we have that R is (α, S) strong, but R does not witness that $\mathcal{J}_\alpha^R$ is S-sound.

In view of the fact that $K(S)$ is independent of S, one might expect the same to be true of (α, S)-strength. This is indeed the case.

Lemma 6.3. *Suppose $K(S)$ and $K(T)$ exist, and $\alpha \leq OR \cap K(S) \cap K(T)$; then for any $\mathcal{M}$, $\mathcal{M}$ is (α, S) strong iff $\mathcal{M}$ is (α, T) strong.*

Proof. Suppose $\mathcal{M}$ is (α, S)-strong. Let $\mathcal{R}$ witness that $\mathcal{J}_\alpha^{\mathcal{M}}$ is S-sound, and W witness that $\mathcal{J}_\alpha^{\mathcal{M}}$ is T-sound. Let Σ be an $(\omega, \Omega + 1)$ iteration strategy for W, and Γ an $(\omega, \Omega + 1)$ iteration strategy for R. From the proof of 5.16, we get iteration trees $\mathcal{T}$ and $\mathcal{U}$ on W and R which are plays of two rounds of $\mathcal{G}^*(W, (\omega, \Omega + 1))$ and $\mathcal{G}^*(R, (\omega, \Omega + 1))$ according to Σ and Γ respectively, and such that $\mathcal{T}$ and $\mathcal{U}$ have a common last model Q. The proof of 5.16 also shows that the iteration maps $\sigma : W \rightarrow Q$ and $\tau : R \rightarrow Q$ satisfy $\alpha \leq \min(\text{crit}(\sigma), \text{crit}(\tau))$. Since $\alpha \leq \text{crit}(\sigma)$, $\nu(E_\gamma^{\mathcal{T}}) \geq \alpha$ for all $\gamma + 1 < lh\ \mathcal{T}$.

Now Σ yields an $(\omega, \Omega + 1)$-iteration strategy Σ^* for Q, and the strategy of copying via τ and using Σ^* on the copied tree is an $(\omega, \Omega + 1)$-iteration strategy for R; call it Σ^{**}.
According to 6.1, there is an iteration tree $\mathcal{V}$ on R having last model $\mathcal{P}$ which is a play by Σ^{**}, and such that $\forall\gamma(\gamma + 1 < lh\ \mathcal{V} \Rightarrow \nu(E_\gamma^{\mathcal{V}}) \geq \alpha)$, and an embedding $\pi : \mathcal{M} \rightarrow \mathcal{P}'$ for some $\mathcal{P}' \trianglelefteq \mathcal{P}$ such that $\pi \restriction \alpha = \text{identity}$. Let $\tau^* : \mathcal{P} \rightarrow \mathcal{L}$, where $\mathcal{L}$ is the last model of the copied tree $\tau\mathcal{V}$ on Q, be the copy map; thus $\tau^* \restriction \alpha = \tau \restriction \alpha = \text{identity}$. Let $\mathcal{L}' \trianglelefteq \mathcal{L}$ correspond to $\mathcal{P}'$. Then $\mathcal{L}'$ is an initial segment of the last model of $\mathcal{T}^\frown\tau\mathcal{V}$, which is a play by Σ; moreover $\tau^* \circ \pi$ maps $\mathcal{M}$ into $\mathcal{L}'$ and $(\tau^* \circ \pi) \restriction \alpha = \text{identity}$.

This shows that $\mathcal{M}$ is (α, T)-strong, as desired. □

Definition 6.4. *Let $\mathcal{M}$ be a proper premouse, and let $\alpha < \Omega$. We say $\mathcal{M}$ is α-strong iff for some S, $\mathcal{M}$ is (α, S)-strong.*

We proceed to the inductive definition of "α-strong". The definition is based on a certain iterability property: roughly speaking, $\mathcal{M}$ is α-strong just in case $\mathcal{M}$ is jointly iterable with any $\mathcal{N}$ which is β-strong for all $\beta < \alpha$. In order to describe this iterability property we must introduce iteration trees whose "base" is not a single model, but rather a family of models. Such systems were called "psuedo-iteration trees" in [FSIT]. Here we shall simply call them iteration trees, and distinguish them from the iteration trees considered so far by means of their bases.

Definition 6.5. *A* simple phalanx *is a pair $(\langle \mathcal{M}_\beta \mid \beta \leq \alpha \rangle, \langle \lambda_\beta \mid \beta < \alpha \rangle)$ such that for all $\beta \leq \alpha$, $\mathcal{M}_\beta$ is an ω-sound proper premouse, and*

(1) $\beta \leq \gamma \leq \alpha \Rightarrow (\mathcal{M}_\gamma \models$ *"λ_β is a cardinal" and* $\rho_\omega(\mathcal{M}_\gamma) \geq \lambda_\beta)$,

(2) $\beta < \gamma \leq \alpha \Rightarrow \mathcal{M}_\beta$ *agrees with* $\mathcal{M}_\gamma$ *below* λ_β, *and*

(3) $\beta < \gamma < \alpha \Rightarrow \lambda_\beta < \lambda_\gamma$.

We have added the qualifier "simple" in 6.5 because we shall introduce a more general kind of phalanx in §9. Since we shall consider only simple phalanxes in this section, we shall drop the "simple" when referring to them.

If $\mathcal{B} = (\langle \mathcal{M}_\beta \mid \beta \leq \alpha \rangle, \langle \lambda_\beta \mid \beta < \alpha \rangle)$ is a phalanx, then we set $lh\ \mathcal{B} = \alpha + 1$, $\mathcal{M}_\beta^{\mathcal{B}} = \mathcal{M}_\beta$ for $\beta \leq \alpha$, and $\lambda(\beta, \mathcal{B}) = \lambda_\beta$ for $\beta < \alpha$.

A phalanx of length 1 is just a premouse. Iteration trees on phalanxes are the obvious generalization of iteration trees on premice; the main point is that we use $\lambda(\beta, \mathcal{B})$ to tell us when to apply an extender to $\mathcal{M}_\beta^{\mathcal{B}}$, just as we used $\nu(E_\beta^{\mathcal{T}})$ in the special case of a tree on a premouse. We shall have $\beta T \gamma$ for $\beta < \gamma < lh\ \mathcal{B}$, but this is only a notational convenience, and it would be more natural to think of a tree with $lh\ \mathcal{B}$ many roots. Since we only need normal, ω-maximal trees, we shall only define these.

Definition 6.6. *Let $\mathcal{B}$ be a phalanx of length $\alpha + 1$, and $\theta > \alpha + 1$. An (ω-maximal, normal) iteration tree of length θ on $\mathcal{B}$ is a system $\mathcal{T} = \langle E_\beta \mid \alpha+1 \leq \beta + 1 < \theta \rangle$ with associated tree order T, models $\mathcal{M}_\beta$ for $\beta < \theta$, and $D \subseteq \theta$ and embeddings $i_{\eta\beta} : \mathcal{M}_\eta \to \mathcal{M}_\beta$ defined for $\eta T \beta$ with $(\alpha \cup D) \cap (\eta, \beta]_T = \emptyset$, such that*

(1) *$\mathcal{M}_\beta = \mathcal{M}_\beta^{\mathcal{B}}$ for all $\beta \leq \alpha$, and for β, $\gamma \leq \alpha$, $\beta T \gamma$ iff $\beta < \gamma$;*

(2) *$\forall \beta < \alpha(\lambda(\beta, \mathcal{B}) < lh\ E_\alpha)$, and for $\alpha + 1 \leq \beta + 1 < \gamma + 1 < \theta$, $lh\ E_\beta < lh\ E_\gamma$;*

(3) *for $\alpha + 1 \leq \beta + 1 < \theta$: T-pred $(\beta + 1)$ is the least ordinal γ such that $\gamma < \alpha$ and $crit(E_\beta) < \lambda(\gamma, \mathcal{B})$, or $\alpha \leq \gamma$ and $crit(E_\beta) < \nu(E_\alpha)$. Moreover, letting $\gamma = T$-pred $(\beta + 1)$ and $\kappa = crit(E_\beta)$,*

$$\mathcal{M}_{\beta+1} = Ult_k(\mathcal{M}_\gamma^*, E_\beta^{\mathcal{T}}),$$

where $\mathcal{M}_\gamma^*$ is the longest initial segment of $\mathcal{M}_\gamma$ containing only subsets of κ measured by E_β, and k is largest such that $\kappa < \rho_k(\mathcal{M}_\gamma^*)$. Also, $\beta + 1 \in D$ iff $\mathcal{M}_\gamma \neq \mathcal{M}_\gamma^*$, and if $\beta + 1 \notin D$ then $i_{\gamma,\beta+1}$ is the canonical embedding from $\mathcal{M}_\gamma$ into $Ult_k(\mathcal{M}_\gamma, E_\beta)$, and $i_{\eta,\beta+1} = i_{\gamma,\beta+1} \circ i_{\eta\gamma}$ for $\eta T \gamma$ such that $D \cap (\eta, \gamma]_T = \emptyset$;

(4) *if $\alpha < \beta < \theta$ and β is a limit, then $D \cap [0, \beta)_T$ is finite, $[0, \beta)_T$ is cofinal in β, and $\mathcal{M}_\beta$ is the direct limit of the $\mathcal{M}_\gamma$ for $\gamma \in [0, \beta)_T$ such that $\gamma \geq \alpha \cup \sup(D)$. Moreover, $i_{\gamma\beta} : \mathcal{M}_\gamma \to \mathcal{M}_\beta$ is the direct limit map for all $\gamma \geq \alpha \cup \sup(D)$.*

In the situation of 6.6, we set $\theta = lh\ \mathcal{T}$, $\mathcal{M}_\beta = \mathcal{M}_\beta^{\mathcal{T}}$, $E_\beta = E_\beta^{\mathcal{T}}$, and so forth. For $\beta < \theta$, we let $\text{root}^{\mathcal{T}}(\beta)$ be the largest $\gamma < lh\ \mathcal{B}$ such that $\gamma T \beta$.

If $\mathcal{B}$ is a phalanx, then $\mathcal{G}^*(\mathcal{B}, \theta)$ is the obvious generalization of the length θ normal iteration game on premice: I and II build an iteration tree on $\mathcal{B}$, with I extending the tree at successor steps and II at limit steps. If at some move $\alpha < \theta$, I produces an illfounded ultrapower or II does not play a cofinal wellfounded branch, then I wins, and otherwise II wins. A winning strategy for II in $\mathcal{G}^*(\mathcal{B}, \theta)$ is a *θ-iteration strategy for $\mathcal{B}$*, and $\mathcal{B}$ is *θ-iterable* just in case there is such a strategy.

We wish to state an iterability theorem for phalanxes which are generated from iterates of K^c.

Definition 6.7. *Let $\mathcal{R}$ be a proper premouse and Σ an $(\omega, \Omega + 1)$ iteration strategy for $\mathcal{R}$. We say that a phalanx $\mathcal{B}$ is $(\Sigma, \mathcal{R})$-generated iff for* *all $\beta < lh\ \mathcal{B}$, there is an almost normal iteration tree $\mathcal{T}$ on $\mathcal{R}$ which is a play according to Σ such that $\mathcal{M}_\beta \trianglelefteq \mathcal{P}$, where $\mathcal{P}$ is the last model of $\mathcal{T}$, and such that (i) if $\beta + 1 < lh\ \mathcal{B}$, then $\lambda(\beta, \mathcal{B})$ is a cardinal of $\mathcal{R}$ and $\forall\gamma\ (\gamma + 1 < lh\ \mathcal{T} \Rightarrow \nu(E_\gamma^{\mathcal{T}}) \geq \lambda(\beta, \mathcal{B}))$, and (ii) if $\beta + 1 = lh\ \mathcal{B}$, then $\forall\gamma\ \forall\alpha < \beta(\gamma + 1 < lh\ \mathcal{T} \Rightarrow \nu(E_\gamma^{\mathcal{T}}) \geq \lambda(\alpha, \mathcal{B}))$.*

Recall that if K^c has no Woodin cardinals, then there is a unique $(\omega, \Omega+1)$ iteration strategy for K^c (namely, choosing the unique cofinal wellfounded branch).

Definition 6.8. *Suppose $K^c \models$ "There are no Woodin cardinals"; then a phalanx $\mathcal{B}$ is K^c-generated iff $\mathcal{B}$ is (Σ, K^c) generated, where Σ is the unique $(\omega, \Omega + 1)$ iteration strategy for K^c.*

Our iterability proof for K^c in §9 will actually show:

Theorem 6.9. *Suppose $K^c \models$ "There are no Woodin cardinals"; then every K^c-generated phalanx $\mathcal{B}$ such that $lh\ \mathcal{B} < \Omega$ is $\Omega + 1$-iterable.*

Proof. Deferred to §9. □

We shall actually only characterize α strength inductively in the case α is a cardinal of K. In this case we have the following little lemma.

Lemma 6.10. *Suppose $K^c \models$"There are no Woodin cardinals", and let α be a cardinal of K. Suppose $\alpha < OR^{\mathcal{M}}$, and $\mathcal{M}$ is α strong. Then α is a cardinal of $\mathcal{M}$.*

Proof. There is a weasel W which witnesses that $\mathcal{J}_\alpha^W = \mathcal{J}_\alpha^K$ is S-sound, and an elementary $\pi : K \to W$ with $\text{crit}(\pi) \geq \alpha$. Since α is a cardinal of K, α is a cardinal of W. But then α is a cardinal of $\mathcal{P}$, whenever $\mathcal{P}$ is an initial segment of a model on an iteration tree $\mathcal{T}$ on W such that $lh(E_\gamma^{\mathcal{T}}) \geq \alpha$ for all $\gamma + 1 < lh\ \mathcal{T}$. We have $\sigma : \mathcal{M} \to \mathcal{P}$ with $\text{crit}(\sigma) \geq \alpha$, for some such $\mathcal{P}$, and this implies that α is a cardinal of $\mathcal{M}$. □

We can now prove the main result of this section.

Theorem 6.11. *Suppose K^c has no Woodin cardinals. Let $\mathcal{M}$ be a proper premouse, and let $\alpha < OR^{\mathcal{M}}$ be such that α is a cardinal of K and $\mathcal{J}_\alpha^{\mathcal{M}} = \mathcal{J}_\alpha^K$; then the following are equivalent:*

(1) $\mathcal{M}$ is α strong,

(2) if $((\mathcal{N}, \mathcal{M}), (\alpha))$ is a phalanx such that $\mathcal{N}$ is β strong for all K-cardinals $\beta < \alpha$, then $((\mathcal{N}, \mathcal{M}), (\alpha))$ is $\Omega + 1$ iterable.

Proof. We show first (2)⇒(1). Let W witness that $\mathcal{J}_\alpha^{\mathcal{M}}$ is S-sound, and let Σ be an $\Omega+1$ iteration strategy for W. By 6.2, W is β strong for all $\beta < \alpha$, and so our hypothesis (2) gives us an $\Omega + 1$ iteration strategy Γ for the phalanx $((W, \mathcal{M}), (\alpha))$. We now compare $\mathcal{M}$ with W, using Σ to form an iteration tree $\mathcal{T}$ on W and Γ to form an iteration tree $\mathcal{U}$ on $((W, \mathcal{M}), (\alpha))$. The trees $\mathcal{T}$

and $\mathcal{U}$ are determined by iterating the least disagreement, starting from $\mathcal{M}$ vs. W, as well as by the rules for iteration trees and the iteration strategies. (See 8.1 of [FSIT] for an example of such a coiteration.)

Let $lh\ \mathcal{U} = \theta + 1$ and $lh\ \mathcal{T} = \gamma + 1$. We claim that $\text{root}^{\mathcal{U}}(\theta) = 1$. For otherwise $\text{root}^{\mathcal{U}}(\theta) = 0$, and the universality of W implies that $\mathcal{M}^{\mathcal{U}}_{\theta} = \mathcal{M}^{\mathcal{T}}_{\gamma}$, and that $i^{\mathcal{U}}_{0\theta}$ and $i^{\mathcal{T}}_{0\gamma}$ exist. Moreover, the rules for $\mathcal{U}$ guarantee that $\text{crit}(i^{\mathcal{U}}_{0\theta}) < \alpha$. Since W has the S-hull and S-definability properties at all $\beta < \alpha$, we then get the usual contradiction involving the common fixed points of $i^{\mathcal{U}}_{0\theta}$ and $i^{\mathcal{T}}_{0\gamma}$.

Thus $\text{root}^{\mathcal{U}}(\theta) = 1$. Since W is universal, $i^{\mathcal{U}}_{1,\theta}$ exists, and maps $\mathcal{M}$ into some initial segment of $\mathcal{M}^{\mathcal{T}}_{\gamma}$. By the rules for $\mathcal{U}$, $\text{crit}(i^{\mathcal{U}}_{1,\theta}) \geq \alpha$. Thus $\mathcal{T}$ and $i^{\mathcal{U}}_{1,\theta}$ witness that $\mathcal{M}$ is (α, S) strong.

We now prove (1)$\Rightarrow$(2). Let us consider first the case α is a successor cardinal of K, say $\alpha = (\beta^{+})^{K} = (\beta^{+})^{\mathcal{M}}$ where β is a cardinal of K. Let $(\langle \mathcal{N}, \mathcal{M} \rangle, \alpha)$ be a phalanx such that $\mathcal{N}$ is β-strong. We shall show $(\langle \mathcal{N}, \mathcal{M} \rangle, \alpha)$ is $\Omega{+}1$ iterable by embedding it into a K^{c}-generated phalanx, and then using 6.9.

Note that $\mathcal{M}$ and $\mathcal{N}$ agree below α, and since $\mathcal{M}$ is α- strong, $\mathcal{J}^{\mathcal{M}}_{\alpha}$ is A_0-sound. Let W be a weasel which witnesses that $\mathcal{J}^{\mathcal{M}}_{\alpha}$ is A_0-sound. By Definition 6.1, there are (finite compositions of normal) iteration trees $\mathcal{T}_0$ and $\mathcal{T}_1$ on W, having last models $\mathcal{P}_0$ and $\mathcal{P}_1$ respectively, such that $\forall \gamma[(\gamma + 1 < lh\ \mathcal{T}_0 \Rightarrow \nu(E^{\mathcal{T}_0}_{\gamma}) \geq \beta)$ and $(\gamma + 1 < lh\ \mathcal{T}_0 \Rightarrow \nu(E^{\mathcal{T}_1}_{\gamma}) \geq \alpha)]$, and there are fully elementary embeddings τ_0 and τ_1 such that

$$\tau_0 : \mathcal{N} \to \mathcal{J}^{\mathcal{P}_0}_{\eta_0} \quad \text{and} \quad \tau_0 \restriction \beta = \text{identity}\,,$$

and

$$\tau_1 : \mathcal{M} \to \mathcal{J}^{\mathcal{P}_1}_{\eta_1} \quad \text{and} \quad \tau_1 \restriction \alpha = \text{identity}\,.$$

The proof of 5.10 shows that we may assume our A_0-soundness witness W is chosen so that there is an elementary $\sigma : W \to K^{c}$. Since α is a cardinal of K, we may also assume that α is a cardinal of W. Let $\sigma\mathcal{T}_0$ and $\sigma\mathcal{T}_1$ be the copied versions of $\mathcal{T}_0$ and $\mathcal{T}_1$ on K^{c}. Since W has no Woodin cardinals (because K^{c} has none), the trees $\mathcal{T}_0$ and $\mathcal{T}_1$ are simple. This implies that the copying construction does not break down, and that $\sigma\mathcal{T}_0$ and $\sigma\mathcal{T}_1$ are according to the unique $(\omega, \Omega + 1)$ iteration strategy for K^{c}. If E is an extender used in $\sigma\mathcal{T}_0$, then $\nu(E) \geq \sigma(\beta)$, and if E is used in $\sigma\mathcal{T}_1$, then $\nu(E) \geq \sigma(\alpha)$. Let

$$\psi_0 : \mathcal{P}_0 \to \mathcal{Q}_0 \quad \text{and} \quad \psi_1 : \mathcal{P}_1 \to \mathcal{Q}_1$$

be the copy maps, where $\mathcal{Q}_0$ and $\mathcal{Q}_1$ are the last models of $\sigma\mathcal{T}_0$ and $\sigma\mathcal{T}_1$ respectively. We have $\psi_0 \restriction \beta = \sigma \restriction \beta$ and $\psi_1 \restriction \alpha = \sigma \restriction \alpha$. Let, for $i \in \{0, 1\}$,

$$\mathcal{R}_i = \begin{cases} \mathcal{Q}_i & \text{if} \quad \mathcal{P}_i = \mathcal{J}^{\mathcal{P}_i}_{\eta_i}\,, \\ \mathcal{J}^{\mathcal{Q}_i}_{\psi_i(\eta_i)} & \text{otherwise}\,. \end{cases}$$

We claim that $(\langle \mathcal{R}_0, \mathcal{R}_1 \rangle, \langle \sigma(\alpha) \rangle)$ is a K^{c}-generated phalanx, the trees by which it is generated being $\sigma\mathcal{T}_0$ and $\sigma\mathcal{T}_1$. For this, we must look more closely at the extenders used in $\mathcal{T}_0$. We claim that if E is used in $\mathcal{T}_0$, then $lh\ E > \alpha$. For if some E such that $lh\ E < \alpha$ is used in $\mathcal{T}_0$, then there is a $B \subseteq \beta$ such that $B \in \mathcal{J}_\alpha^W$ and $B \notin \mathcal{P}_0$. Since $\mathcal{M}, \mathcal{N}$, and W agree below α, $B \in \mathcal{N}$, so $\tau_0(B) \in \mathcal{P}_0$, so $\tau_0(B) \cap \beta = B \in \mathcal{P}_0$, a contradiction. Also, $lh\ E \neq \alpha$ for all E on the W sequence, since α is a cardinal of W. Thus $lh\ E > \alpha$ for all E used in $\mathcal{T}_0$. Since α is a cardinal of W, this means $\nu(E) \geq \alpha$ for all E used in $\mathcal{T}_0$. That implies that $\nu(E) \geq \sigma(\alpha)$ for all E used in $\sigma\mathcal{T}_0$. The remaining clauses in the definition of "K^c-generated phalanx" hold obviously of $(\langle \mathcal{R}_0, \mathcal{R}_1\rangle, \langle \sigma(\alpha)\rangle)$.

By 6.9 we have an $\Omega + 1$ iteration strategy Σ for $(\langle \mathcal{R}_0, \mathcal{R}_1\rangle, \langle \sigma(\alpha)\rangle)$. We can use Σ and a simple copying construction to get an $\Omega+1$ iteration strategy Γ for $(\langle \mathcal{N}, \mathcal{M}\rangle, \langle \alpha\rangle)$. We shall describe this construction now; it involves a small wrinkle on the usual copying procedure, and it shows why it is necessary that $\mathcal{M}$ be α-strong, and not just β-strong.

Our strategy Γ is to insure that if $\mathcal{T}$ is the iteration tree on $(\langle \mathcal{N}, \mathcal{M}\rangle, \langle \alpha\rangle)$ representing the current position in $\mathcal{G}^*((\langle \mathcal{N}, \mathcal{M}\rangle, \langle \alpha\rangle), \Omega+1)$, then as we built $\mathcal{T}$ we constructed an iteration tree $\mathcal{U}$ on $(\langle \mathcal{R}_0, \mathcal{R}_1\rangle, \langle \sigma(\alpha)\rangle)$ such that $\mathcal{U}$ is a play by Σ and has the same tree order as $\mathcal{T}$, together with embeddings

$$\pi_\gamma : \mathcal{M}_\gamma^{\mathcal{T}} \to \mathcal{M}_\gamma^{\mathcal{U}}$$

defined for all $\gamma < lh\ \mathcal{T}$, satisfying:

(a) for $\eta < \gamma < lh\ \mathcal{T}$, $\pi_\eta \restriction \nu_\eta = \pi_\gamma \restriction \nu_\eta$, where

$$\nu_\eta = \begin{cases} \beta & \text{if } \eta = 0, \\ \nu(E_\eta^{\mathcal{T}}) & \text{if } \eta > 0 \text{ and } E_\eta^{\mathcal{T}} \text{ is of type III}, \\ lh(E_\eta^{\mathcal{T}}) & \text{otherwise}; \end{cases}$$

moreover, $E_\eta^{\mathcal{U}} = \pi_\eta(E_\eta^{\mathcal{T}})$;

(b) for all $\gamma < lh\ \mathcal{T}$ such that $\gamma \geq 2$, π_γ is a $(\deg^{\mathcal{T}}(\gamma), X)$ embedding, where $X = (i_{\eta\gamma}^{\mathcal{T}} \circ i_\eta^*)''(\mathcal{M}_\eta^*)^{\mathcal{T}}$, for η the least ordinal such that $i_{\eta\gamma}^{\mathcal{T}} \circ i_\eta^*$ exists; for $\gamma \in \{0, 1\}$, π_γ is fully elementary;

(c) for $\eta < \gamma < lh\ \mathcal{T}$, if $i_{\eta\gamma}^{\mathcal{T}}$ exists, then $i_{\eta\gamma}^{\mathcal{U}}$ exists and $\pi_\gamma \circ i_{\eta\gamma}^{\mathcal{T}} = i_{\eta\gamma}^{\mathcal{U}} \circ \pi_\eta$.

These are just the usual copying conditions, except that the agreement-of-embeddings ordinal ν_0 is β, rather than α.

We have $\mathcal{M}_0^{\mathcal{T}} = \mathcal{N}$, $\mathcal{M}_1^{\mathcal{T}} = \mathcal{M}$, $\mathcal{M}_0^{\mathcal{U}} = \mathcal{R}_0$, and $\mathcal{M}_1^{\mathcal{U}} = \mathcal{R}_1$ to begin with, and we set

$$\pi_0 = \psi_0 \circ \tau_0 \text{ and } \pi_1 = \psi_1 \circ \tau_1 .$$

Since $\pi_0 \restriction \beta = \pi_1 \restriction \beta$ and π_0, π_1 are fully elementary, our induction hypotheses (a) - (d) hold.

[To see π_0 and π_1 are fully elementary, notice that $\mathcal{M}$ and $\mathcal{N}$ satisfy ZF-Powerset, and τ_0 and τ_1 are fully elementary according to 6.1. If $\mathcal{J}_{\eta_i}^{\mathcal{P}_i} = \mathcal{P}_i$, this means $\mathcal{P}_i \models$ ZF-Powerset, so $\deg^{\mathcal{T}_i}(\xi_i) = \omega$, where $\mathcal{P}_i = \mathcal{M}_{\xi_i}^{\mathcal{T}_i}$, and thus ψ_i is fully elementary ($i \in \{0, 1\}$). On the other hand, if $\mathcal{J}_{\eta_i}^{\mathcal{P}_i}$ is a proper initial segment of $\mathcal{P}_i$, then $\psi_i \restriction \mathcal{J}^{\mathcal{P}_i}_{\eta_i}$ is obviously fully elementary. So in any case π_0 and π_1 are fully elementary.]

Now suppose we are at a limit step λ in the construction of $\mathcal{T}$ and $\mathcal{U}$. Σ chooses a cofinal wellfounded branch b of $\mathcal{U} \restriction \lambda$, and we let Γ choose b as its cofinal wellfounded branch of $\mathcal{T} \restriction \lambda$. It is cofinal because $\mathcal{T}$ and $\mathcal{U}$ have the same tree order, and wellfounded because we have an embedding $\pi : \mathcal{M}^{\mathcal{T}}_b \to \mathcal{M}^{\mathcal{U}}_b$ given by

$$\pi(i^{\mathcal{T}}_{\gamma b}(x)) = i^{\mathcal{U}}_{\gamma b}(\pi_\gamma(x))$$

defined for $\gamma \in b$ sufficiently large. Setting $\pi_\lambda = \pi$, we can easily check (a) - (d).

Now suppose we are at step $\eta + 1$ in the construction of $\mathcal{T}$ and $\mathcal{U}$. Player I in $\mathcal{G}^*(\langle \mathcal{N}, \mathcal{M} \rangle, \langle \alpha \rangle)$ has just played $E^{\mathcal{T}}_\eta$, and thereby determined $\mathcal{T} \restriction \eta + 2$. We must determine $\mathcal{U} \restriction \eta + 2$ together with $\pi_{\eta+1}$. In the case that T-pred$(\eta + 1) \neq 0$, we can simply quote the shift lemma, Lemma 5.2 of [FSIT], and obtain the desired $\mathcal{M}^{\mathcal{U}}_{\eta+1}$ and $\pi_{\eta+1}$. We omit further detail, and go on to the case T-pred$(\eta + 1) = 0$. [Unfortunately, the agreement-of-embeddings hypothesis for the copying construction was mis-stated in [FSIT], because squashed ultrapowers were overlooked. We only get $\pi_\eta \restriction \nu(E^{\mathcal{T}}_\eta) = \pi_\gamma \restriction \nu(E^{\mathcal{T}}_\eta)$, for $\eta < \gamma$, in the case $E^{\mathcal{T}}_\eta$ is type III, rather than $\pi_\eta \restriction (lh(E^{\mathcal{T}}_\eta) + 1) = \pi_\gamma \restriction (lh\ E^{\mathcal{T}}_\eta + 1)$ as claimed in [FSIT] (after 5.2, in the definition of $\pi \mathcal{T}$). This weaker agreement causes no new problems, however.]

Let $\kappa = \text{crit}(E^{\mathcal{T}}_\eta)$, so that $\kappa < \alpha$ and hence $\kappa \leq \beta$. To simply quote the Shift lemma we would need that $\pi_0 \restriction (\kappa^+)^{\mathcal{M}^{\mathcal{T}}_\eta} = \pi_\eta \restriction (\kappa^+)^{\mathcal{M}^{\mathcal{T}}_\eta}$, and that is more than we know. Still, the proof of the Shift lemma works: set

$$\mathcal{M}^{\mathcal{U}}_{\eta+1} = \text{Ult}_\omega(\mathcal{R}_0, \pi_\eta(E^{\mathcal{T}}_\eta)).$$

From 6.6 (2), we get $\nu_1 \geq \alpha$, and our agreement hypothesis (a) then gives $\pi_1 \restriction \alpha = \pi_\eta \restriction \alpha$. Thus $\pi_\eta(\kappa) = \pi_1(\kappa) < \pi_1(\alpha)$. Also, $\pi_1(\alpha) = \sigma(\alpha)$. (Since $\tau_1 \restriction \alpha =$ identity and $\psi_1 \restriction \alpha = \sigma \restriction \alpha$, $\pi_1(\beta) = \sigma(\beta)$. But $\pi_1(\alpha)$ is the $\mathcal{R}_1$-successor cardinal of $\pi_1(\beta)$, and $\sigma(\alpha)$ is the K^c-successor cardinal of $\sigma(\beta)$, and since all extenders used in $\sigma\mathcal{T}_1$ have length $> \sigma(\alpha)$, these are the same.) Since $\mathcal{M}^{\mathcal{U}}_\eta$ agrees with $\mathcal{R}_1$, and hence $\mathcal{R}_0$, through $\sigma(\alpha) = \pi_1(\alpha)$, $\mathcal{M}^{\mathcal{U}}_\eta$ and $\mathcal{R}_0$ have the same subsets of $\pi_\eta(\kappa)$, and the ultrapower defining $\mathcal{M}^{\mathcal{U}}_{\eta+1}$ makes sense.

We can now define $\pi_{\eta+1} : \mathcal{M}^{\mathcal{T}}_{\eta+1} \to \mathcal{M}^{\mathcal{U}}_{\eta+1}$ by:

$$\pi_{\eta+1}([a, f]^{\mathcal{T}}_{E^{\mathcal{T}}_\eta}) = [\pi_\eta(a), \pi_0(f) \restriction [\pi_\eta(\kappa)]^{|a|}]^{\mathcal{R}_0}_{\pi_\eta(E^{\mathcal{T}}_\eta)}.$$

The shift lemma argument shows that $\pi_{\eta+1}$ is well defined, fully elementary, and has the desired agreement with π_η. To see this, recall that $\nu(E) \geq \alpha$ for all E used in $\sigma\mathcal{T}_0$. This implies that $\psi_0 \restriction \alpha = \sigma \restriction \alpha$, and thus ψ_0, ψ_1, π_1, and π_η all agree with σ on α. Now $\kappa \leq \beta$, and for any $A \subseteq \beta$ in $\mathcal{N}$,

$$\begin{aligned}\pi_0(A)\cap\pi_\eta(\beta) &= \pi_0(A)\cap\psi_0(\beta)\\ &= \psi_0(\tau_0(A))\cap\psi_0(\beta)\\ &= \psi_0(\tau_0(A)\cap\beta)\\ &= \psi_0(A)\\ &= \pi_\eta(A)\,.\end{aligned}$$

Thus, for example, if $f = g$ on $A \subseteq [\kappa]^{|a|}$ with $A \in (E_\eta^{\mathcal{T}})_a$, then $\pi_0(f) = \pi_0(g)$ on $\pi_0(A)$, and hence $\pi_0(f) = \pi_0(g)$ on $\pi_0(A) \cap [\pi_\eta(\kappa)]^{|a|}$. But then $\pi_0(f) = \pi_0(g)$ on $\pi_\eta(A)$, and $\pi_\eta(A) \in (\pi_\eta(E_\eta^{\mathcal{T}}))_{\pi_\eta(a)}$. This shows that $\pi_{\eta+1}$ is well defined, and the other conditions on it can also be checked easily.

This completes the proof of (1)$\Rightarrow$(2) in the case that α is a successor cardinal of K. It is worth noting that we really used that $\mathcal{M}$ was α-strong, and not just β-strong. This guaranteed $\tau_1 \restriction \alpha = \mathrm{id}$, and thus $\pi_1 \restriction \alpha = \sigma \restriction \alpha$. That in turn gave $\pi_\eta \restriction \alpha = \psi_0 \restriction \alpha$, which was crucial. It is not true that if $\mathcal{M}$ is β-strong, where β is a cardinal of K, and $\mathcal{J}_\alpha^{\mathcal{M}} = \mathcal{J}_\alpha^K$ for $\alpha = (\beta^+)^K$, then $\mathcal{M}$ is α-strong.

The case α is a limit cardinal of K is similar. Let $\mathcal{N}$ be β-strong for all K-cardinals $\beta < \alpha$, and $\mathcal{J}_\alpha^{\mathcal{N}} = \mathcal{J}_\alpha^{\mathcal{M}}$. Let W witness that $\mathcal{J}_\alpha^{\mathcal{M}} = \mathcal{J}_\alpha^K$ is A_0-sound, and let $\sigma : W \to K^c$. For each K-cardinal $\beta \leq \alpha$ let $\mathcal{T}_\beta$ be an iteration tree on W with last model $\mathcal{P}_\beta$, and let $\tau_\beta : \mathcal{N} \to \mathcal{J}_{\eta_\beta}^{\mathcal{P}_\beta}$ with $\tau_\beta \restriction \beta = \mathrm{id}$ for $\beta < \alpha$. Let $\tau_\alpha : \mathcal{M} \to \mathcal{J}_{\eta_\alpha}^{\mathcal{P}_\alpha}$ with $\tau_\alpha \restriction \alpha = \mathrm{id}$. For $\beta \leq \alpha$, let $\sigma\mathcal{T}_\beta$ be the copied tree on K^c, Q_β its last model, $\psi_\beta : \mathcal{P}_\beta \to Q_\beta$ the copy map, and $\mathcal{R}_\beta = \mathcal{J}_{\psi_\beta(\eta_\beta)}^{Q_\beta}$ or $\mathcal{R}_\beta = Q_\beta$ as appropriate. Then $(\langle \mathcal{R}_\beta \mid \beta \leq \alpha \wedge \beta$ a cardinal of $K\rangle, \langle \sigma(\beta) \mid \beta < \alpha \wedge \beta$ a cardinal of $K\rangle)$ is a K^c-generated phalanx, and therefore $\Omega + 1$ iterable. But then we can win the iteration game $\mathcal{G}^*((\langle \mathcal{N}, \mathcal{M}\rangle, \langle \alpha \rangle), \Omega + 1)$ just as before; letting $\pi_\beta : \mathcal{N} \to \mathcal{T}_\beta$ be given by $\pi_\beta = \psi_\beta \circ \tau_\beta$, for $\beta \leq \alpha$, and defining the remaining π's inductively, we copy the evolving $\mathcal{T}$ on $(\langle \mathcal{N}, \mathcal{M}\rangle, \langle \alpha \rangle)$ by applying $\pi_\eta(E_\eta^{\mathcal{T}})$ to the model required by the rules for trees on $(\langle \mathcal{R}_\beta \mid \beta \leq \alpha \wedge \beta$ a K-cardinal$\rangle, \langle \sigma(\beta) \mid \beta < \alpha \wedge \beta$ a K-cardinal$\rangle)$. Since for $\beta \leq \alpha$, $\pi_\beta \restriction \beta = \psi_\beta \restriction \beta = \sigma \restriction \beta$, we have enough agreement to simply quote the shift lemma. Although $\mathcal{T}$ and its copy $\mathcal{U}$ have slightly different tree orders, this causes no problems.

This completes the proof of 6.11. □

To see that 6.11 gives on inductive definition of K, assuming K^c has no Woodin cardinals, suppose that α is a cardinal of K and we know which premice are α-strong. Then

$$\exists\beta < (\alpha^+)^K(\mathcal{P} = \mathcal{J}_\beta^K) \Leftrightarrow \exists\mathcal{M}(\mathcal{M} \text{ is } \alpha\text{ -strong } \wedge \exists\beta < (\alpha^+)^{\mathcal{M}}(\mathcal{P} = \mathcal{J}_\beta^{\mathcal{M}}))\,.$$

(We get $\Rightarrow$ from 6.2. We get $\Leftarrow$ easily from the definition of "α-strong".)

We can determine $(\alpha^+)^K$ and $\mathcal{J}_{(\alpha^+)^K}^K$ using this equivalence. Using 6.11, we can then determine which premice are $(\alpha^+)^K$-strong. The limit steps in the inductive definition of "α is a cardinal of K" and "$\mathcal{M}$ is α-strong" are trivial modulo 6.11.

This definition still involves quantification over $V_{\Omega+1}$. In order to avoid that, we must show that if $\mathcal{M}$ is of size α, and 6.11 (2) fails, then there is an $\mathcal{N}$ of size α and an iteration tree $\mathcal{T}$ of size α on $(\langle\mathcal{N},\mathcal{M}\rangle,\langle\alpha\rangle)$ witnessing the failure of iterability. (We shall actually get a countable $\mathcal{T}$.) This is a reflection result much like lemma 2.4.

Definition 6.12. *A premouse $\mathcal{M}$ is properly small iff $\mathcal{M} \models$ "There are no Woodin cardinals $\wedge$ there is a largest cardinal". A phalanx $\mathcal{B}$ is properly small iff $\forall\alpha < lh(\mathcal{B})$ ($\mathcal{M}^{\mathcal{B}}_{\alpha}$ is properly small).*

The uniqueness results of §6 of [FSIT] easily yield the following.

Lemma 6.13. *Let $\mathcal{B}$ be a properly small phalanx, and let $\mathcal{T}$ be an iteration tree on $\mathcal{B}$; then $\mathcal{T}$ is simple.*

Proof (Sketch). Suppose b and c are distinct branches of $\mathcal{T}$ with $\sup(b) = \lambda = \sup(c)$, b and c existing in some generic extension of V. If b and c do not drop, then $\delta(\mathcal{T}\restriction\lambda) < \mathrm{OR}^{\mathcal{M}^{\mathcal{T}}_b}$ and $\delta(\mathcal{T}\restriction\lambda) < \mathrm{OR}^{\mathcal{M}^{\mathcal{T}}_c}$ because $\mathcal{M}^{\mathcal{T}}_b$ and $\mathcal{M}^{\mathcal{T}}_c$ have a largest cardinal. (This is why we included this condition.) From 6.1 of [FSIT] we get that $\delta(\mathcal{T}\restriction\lambda)$ is Woodin in $\mathcal{M}^{\mathcal{T}}_b$ if $\mathrm{OR}^{\mathcal{M}^{\mathcal{T}}_b} \leq \mathrm{OR}^{\mathcal{M}^{\mathcal{T}}_c}$, and Woodin in $\mathcal{M}^{\mathcal{T}}_c$ otherwise. This contradicts the proper smallness of the premice in $\mathcal{B}$. If one of b and c drops, then we can argue to a contradiction as in the proof of 6.2 of [FSIT]. □

We thank Kai Hauser for pointing out that our original version of 6.13 was false. (We had omitted having a largest cardinal from the definition of properly small.)

By 6.13, a properly small phalanx can have at most one $\Omega+1$ iteration strategy, that strategy being to choose the unique cofinal wellfounded branch.

Lemma 6.14. *Suppose K^c has no Woodin cardinals, and that α is a cardinal of K. Let $\mathcal{M}$ be a properly small premouse of cardinality α such that $\mathcal{J}^{\mathcal{M}}_{\alpha} = \mathcal{J}^{K}_{\alpha}$ but $\mathcal{M}$ is not α-strong. Then there is a properly small premouse $\mathcal{N}$ of cardinality α such that $\mathcal{J}^{\mathcal{N}}_{\alpha} = \mathcal{J}^{K}_{\alpha}$ and $\mathcal{N}$ α-strong, and a countable putative iteration tree $\mathcal{T}$ on $(\langle\mathcal{N},\mathcal{M}\rangle,\langle\alpha\rangle)$ such that either $\mathcal{T}$ has a last, illfounded model, or $\mathcal{T}$ has limit length but no cofinal wellfounded branch.*

Proof. Let W be a weasel which witnesses that $\mathcal{J}^{K}_{\alpha}$ is A_0-sound. By 6.2, W is α-strong. From the proof of (2)$\Rightarrow$(1) in 6.11, we have that $(\langle W,\mathcal{M}\rangle,\langle\alpha\rangle)$ is not $\Omega+1$ iterable. It follows that there is a putative iteration tree $\mathcal{U}$ of length $\leq\Omega$ on $(\langle W,\mathcal{M}\rangle,\langle\alpha\rangle)$ which is *bad*; i.e., has a last, illfounded model or is of limit length and has no cofinal wellfounded branch.

Since Ω is weakly compact, $lh\ \mathcal{U} < \Omega$. This means that for all sufficiently large successor cardinals μ of W, we can associate to $\mathcal{U}$ a tree $\mathcal{U}_{\mu}$ on $(\langle\mathcal{J}^{W}_{\mu},\mathcal{M}\rangle,\langle\alpha\rangle)$. $\mathcal{U}_{\mu}$ has the same tree order and uses the same extenders as

$\mathcal{U}$; the models on $\mathcal{U}_\mu$ are initial segments of the models on $\mathcal{U}$. We claim that there is a μ such that $\mathcal{U}_\mu$ is bad. If $\mathcal{U}$ has successor length this is obvious, as the last model of $\mathcal{U}$ is the union over μ of the last models of the $\mathcal{U}_\mu$. Suppose $\mathcal{U}$ has limit length, and b_μ is a cofinal wellfounded branch of $\mathcal{U}_\mu$, for all $\mu < \Omega$ such that μ is a successor cardinal of W. Notice that if $\mu < \eta$, then b_η is a cofinal wellfounded branch of $\mathcal{U}_\mu$, and thus by 6.13, $b_\eta = b_\mu$. Letting b be the common value of b_μ for all appropriate μ, we then have that b is a cofinal wellfounded branch of $\mathcal{U}$, a contradiction.

Let $\mathcal{V} = \mathcal{U}_\mu$ and $\mathcal{P} = \mathcal{J}_\mu^W$, where μ is a successor cardinal of W large enough that $\mathcal{V}$ is a bad tree on $(\langle \mathcal{P}, \mathcal{M}\rangle, \langle\alpha\rangle)$. Note that $\mathcal{P}$ is α-strong, and $(\langle \mathcal{P}, \mathcal{M}\rangle, \langle\alpha\rangle)$ is properly small. Let $X \prec V_\eta$, for some η, with $\mathcal{V}, \mathcal{P}, \mathcal{M}, \alpha \in X$, and X countable. Let $\pi : R \cong X$ be the transitive collapse, and $\pi(\bar{\mathcal{V}}) = \mathcal{V}$, etc. Let $\lambda \in X \cap \Omega$ be such that $\mathcal{V}, \mathcal{P}, \mathcal{M}, \alpha \in V_\lambda$; then $V_\lambda^! \in X$, and thus $R \models V_{\bar\lambda}^!$ exists. Because π embeds $(V_{\bar\lambda}^!)^R$ into $V_\lambda^!$, we have $(V_{\bar\lambda}^!)^R = (V_{\bar\lambda}^R)^!$, and so $R[x]$ is correct for Π_2^1 assertions about x, whenever x is an R-generic real coding $V_{\bar\lambda}^R$. But now R satisfies "$\bar{\mathcal{V}}$ is a bad tree on $(\langle \bar{\mathcal{P}}, \bar{\mathcal{M}}\rangle, \langle\bar\alpha\rangle)$", and because $\bar{\mathcal{V}}$ is simple by 6.13, $R[x]$ must satisfy the same. Thus $\bar{\mathcal{V}}$ is indeed a bad tree on $(\langle \bar{\mathcal{P}}, \bar{\mathcal{M}}\rangle, \langle\bar\alpha\rangle)$.

Now let $X \prec Y \prec V_\eta$, where $(\alpha+1) \cup \mathcal{M} \subseteq Y$ and $|Y| \leq \alpha$. Let $\sigma : S \cong Y$ be the transitive collapse, and $\psi : R \to S$ be such that $\pi = \sigma \circ \psi$. Notice that $\psi(\bar{\mathcal{M}}) = \mathcal{M}$ and $\psi(\bar\alpha) = \alpha$. Let $\mathcal{N} = \psi(\bar{\mathcal{P}})$. Using ψ we can copy $\bar{\mathcal{V}}$ as a tree $\psi\bar{\mathcal{V}}$ on $(\langle \mathcal{N}, \mathcal{M}\rangle, \alpha)$, noting that because $\bar{\mathcal{V}}$ is simple, $\psi\bar{\mathcal{V}}$ can never have a wellfounded maximal branch. It follows that $\psi\bar{\mathcal{V}}$ is a bad tree on $(\langle \mathcal{N}, \mathcal{M}\rangle, \langle\alpha\rangle)$. Since $\sigma : \mathcal{N} \to \mathcal{P}$ and $\sigma \restriction (\alpha+1) =$ identity, $\mathcal{N}$ is α-strong. This completes the proof of 6.14. □

Clearly, if α is a cardinal of K and $\beta < (\alpha^+)^K$, then there is a properly small, α-strong $\mathcal{M}$ such that $\mathcal{J}_\beta^{\mathcal{M}} = \mathcal{J}_\beta^K$ and $\beta < (\alpha^+)^{\mathcal{M}}$. So in our inductive definition of K we need only consider properly small mice. Thus 6.11 and 6.14 together yield:

Theorem 6.15. *Suppose K^c has no Woodin cardinals; then there are formulae $\psi(v_0, v_1)$, $\varphi(v_0, v_1)$ in the language of set theory such that whenever G is V-generic/$\mathbb{P}$, where $\mathbb{P} \in V_\Omega$, then $V[G]$ satisfies the following sentences:*

(1) $\forall x, y \in {}^\omega\omega\, \forall\alpha < \omega_1\ [(L_{\alpha+1}(\mathbb{R}) \models \varphi[x,y]) \Leftrightarrow \exists\delta \leq \alpha$ (x codes $\delta \wedge y$ codes a δ-strong, properly small premouse)];

(2) $\forall x, y \in {}^\omega\omega\ \forall\alpha < \omega_1\ [(L_{\alpha+1}(\mathbb{R}) \models \psi[x,y]) \Leftrightarrow \exists\delta \leq \alpha$($x$ codes $\delta \wedge y$ codes $\mathcal{J}_\delta^K$)].

§7. Some applications

In this section, we use the theory developed in §1 - §6 to show that various propositions imply the existence of an inner model with a Woodin cardinal.

A. Saturated ideals

Shelah has shown, in unpublished work, that Con (ZFC + There is a Woodin cardinal) implies Con (ZFC + There is an ω_2-saturated ideal on ω_1). (For earlier results in this direction, see [Ku2], [SVW], [W], and [FMS].) Here we shall prove what is very nearly a converse to Shelah's result. We shall show that Con (ZFC + There is an ω_2-saturated ideal on ω_1+ There is a measurable cardinal) implies Con (ZFC + There is a Woodin cardinal).

The best lower bound on the consistency strength of the existence of an ω_2-saturated ideal on ω_1 known before our work is due to Mitchell ([M?]). He obtained Con (ZFC + $\exists\kappa(o(\kappa) = \kappa^{++})$), which of course is as far as the models studied in [M?] could go.

Actually, our proof does not require that the given ideal be on ω_1, nor does it require ω_2-saturation in full. A generic almost-huge embedding will suffice.

Theorem 7.1. *Let Ω be measurable, and let G be V-generic/ $\mathbb{P}$ for some $\mathbb{P} \in V_\Omega$. Suppose that in $V[G]$ there is a transitive class M and an elementary embedding*

$$j : V \to M \subseteq V[G]$$

with critical point κ such that

$$\forall \alpha < j(\kappa)({}^{\alpha}M \cap V[G] \subseteq M).$$

Then $K^c \models$ There is a Woodin cardinal.

Proof. Suppose toward contradiction that K^c has no Woodin cardinals. This supposition puts the theory of §1- §6 at our disposal. In particular, by 5.18 we have that $K^V = K^{V[G]}$. Moreover, by 6.15, the agreement between M and $V[G]$ implies that if $\mathcal{P}$ is a properly small premouse of cardinality $< j(\kappa)$ in $V[G]$, and $\alpha < j(\kappa)$, then

$$(M \models \mathcal{P} \text{ is } \alpha\text{-strong}) \Leftrightarrow (V[G] \models \mathcal{P} \text{ is } \alpha\text{-strong}).$$

It follows that K^M agrees with $K^{V[G]}$ below $j(\kappa)$. That is, $\mathcal{J}_\alpha^{K^M} = \mathcal{J}_\alpha^{K^{V[G]}}$ for all $\alpha < j(\kappa)$.

Since $\kappa = \text{crit}(j)$, κ is a regular cardinal in V, and thus $j(\kappa)$ is a regular cardinal in M. Since $P(\alpha)^M = P(\alpha)^{V[G]}$ for all $\alpha < j(\kappa)$, $j(\kappa)$ is a cardinal of $V[G]$. Thus $j(\kappa)$ is a cardinal of both K^M and $K^{V[G]}$.

We claim κ is inaccessible in K^V. For otherwise, we have $\beta < \kappa$ such that $\kappa = (\beta^+)^{K^V}$. This means $j(\kappa) = (\beta^+)^{K^M} = (\beta^+)^{K^{V[G]}} = (\beta^+)^{K^V}$, a contradiction. So κ is inaccessible in K^V. But then $j(\kappa)$ is inaccessible in

K^M, and $j(\kappa)$ is a limit cardinal in $K^{V[G]}$. Also, since K^V and K^M agree below $j(\kappa)$, κ is inaccessible in K^M.

Let E_j be the extender over V derived from j. We shall show that for all $\alpha < j(\kappa)$:

$$E_j \cap ([\alpha]^{<\omega} \times K^V) \in K^V .$$

This is a contradiction, as then these fragments of E_j witness that κ is Shelah in K^V. (That is so because $j(K^V) = K^M$ agrees with K^V below $j(\kappa)$.)

So fix $\alpha < j(\kappa)$, and take α large enough that $(\kappa^+)^{K^V} < \alpha$. Let $W \in V$ be a weasel which witnesses that $\mathcal{J}_\alpha^{K^V}$ is A_0-sound. We may assume α is chosen to be a cardinal of K^V and W. It will be enough to find an extender F on the W sequence such that $\text{crit}(F) = \kappa$, $\nu(F) \geq \alpha$, and for all $A \in P(\kappa) \cap K^V$,

$$i_F^W(A) \cap \alpha = j(A) \cap \alpha .$$

Working in $V[G]$, we shall compare W with $j(W)$. Notice that by 5.12, there is in $V[G]$ a (unique) $\Omega + 1$ iteration strategy Σ for W. We shall show that there is a (unique) $\Omega{+}1$ iteration strategy Γ for the phalanx $(\langle W, j(W)\rangle, \langle\alpha\rangle)$. Let us assume for now that such a Γ exists, and complete the proof.

Let $\mathcal{T}$ on W and $\mathcal{U}$ on $(\langle W, j(W)\rangle, \langle\alpha\rangle)$ be the iteration trees resulting from a (Σ, Γ) coiteration. (Coiteration was defined only for premice, but it makes obvious sense for phalanxes. Here we start out comparing $j(W)$ with W, iterating the least disagreement, but the tree $\mathcal{U}$, which begins on $j(W)$, goes back to W whenever it uses an extender with critical point $< \alpha$.) Let $\mathcal{M}_\alpha$ be the αth model of $\mathcal{T}$ and $\mathcal{N}_\alpha$ the αth model of $\mathcal{U}$. In order to save a little notation, let us assume $\mathcal{T}$ and $\mathcal{U}$ are "padded", so that $lh\ \mathcal{T} = lh\ \mathcal{U}$. Let $lh\ \mathcal{T} = lh\ \mathcal{U} = \theta + 1$, where $\theta \leq \Omega$.

We claim that $\text{root}^{\mathcal{U}}(\theta) = 1$. For otherwise, $\text{root}^{\mathcal{U}}(\theta) = 0$; that is, $\mathcal{N}_\theta$ is above $W = \mathcal{N}_0$ in $\mathcal{U}$. Now W is universal, and therefore there is no dropping on $[0, \theta]_U$ or $[0, \theta]_T$, so that $i^{\mathcal{U}}_{0,\theta}$ and $i^{\mathcal{T}}_{0,\theta}$ are defined; moreover, $\mathcal{M}_\theta = \mathcal{N}_\theta$. Let

$$\Delta = \{\gamma < \Omega \mid i^{\mathcal{U}}_{0,\theta}(\gamma) = i^{\mathcal{T}}_{0,\theta}(\gamma) = \gamma\} ,$$

so that Δ is thick in W and $\mathcal{M}_\theta$. The construction of $\mathcal{U}$ guarantees crit $i^{\mathcal{U}}_{0,\theta} < \alpha$. It follows that crit $i^{\mathcal{U}}_{0,\theta}$ is the least γ such that $\gamma \notin H^{\mathcal{M}_\theta}(\Delta)$. From this we get crit $i^{\mathcal{T}}_{0,\theta} =$ crit $i^{\mathcal{U}}_{0,\theta}$. Using the hull property for W at crit $i^{\mathcal{U}}_{0,\theta}$, we proceed to the standard contradiction.

So $\mathcal{N}_\theta$ is above $j(W) = \mathcal{N}_1$ on $\mathcal{U}$. Now $j(W)$ is universal (in $V[G]$) since the class of fixed points of j is α-club in Ω for all sufficiently large regular α, so that $j(W)$ computes α^+ correctly for stationary many $\alpha < \Omega$. Thus $\mathcal{N}_\theta = \mathcal{M}_\theta$, and $i^{\mathcal{U}}_{1,\theta}$ and $i^{\mathcal{T}}_{0,\theta}$ and defined.

Let

$$\Gamma = \{\gamma < \Omega \mid i^{\mathcal{T}}_{0,\theta}(\gamma) = i^{\mathcal{U}}_{1,\theta} \circ j(\gamma) = \gamma\} ,$$

so that Γ is thick in W and $\mathcal{M}_\theta = \mathcal{N}_\theta$. Now $\kappa = \text{crit}(i^{\mathcal{U}}_{1,\theta} \circ j)$, so

$$\kappa = \text{least } \eta \text{ s.t. } \eta \notin H^{\mathcal{N}_\theta}(\Gamma) .$$

It follows that $\kappa = \text{crit } i^{\mathcal{T}}_{0,\theta}$. Similarly, using the hull property for W at κ,

$$i^{\mathcal{T}}_{0,\theta}(A) = i^{\mathcal{U}}_{1,\theta} \circ j(A)$$

for all $A \subseteq \kappa$ s.t. $A \in W$.

Let $\eta+1 \in [0,\theta]_T$ be such that $T\text{-pred}(\eta+1) = 0$. Now all extenders used in $\mathcal{T}$ or $\mathcal{U}$ have length $> \alpha$, and sup of generators $\geq \alpha$. So crit $i^{\mathcal{T}}_{\eta+1,\theta} \geq \alpha$. Also, crit $i^{\mathcal{U}}_{1,\theta} \geq \alpha$ by construction. So for all $A \in P(\kappa)^W$,

$$i^{\mathcal{T}}_{0,\eta+1}(A) \cap \alpha = j(A) \cap \alpha\,.$$

Let F be the trivial completion of $E^{\mathcal{T}}_\eta \restriction \alpha$. Then F is on the sequence of $\mathcal{M}^{\mathcal{T}}_\eta$. It follows (using coherence if $\eta > 0$) that F is on the sequence of $W = \mathcal{M}^{\mathcal{T}}_0$. Moreover, for $A \subseteq \kappa$ s.t. $A \in W$,

$$\begin{aligned} i^{\mathcal{T}}_{0,\eta+1}(A) \cap \alpha &= i^{W}_{E_\eta}(A) \cap \alpha \\ &= i^{W}_{F}(A) \cap \alpha\,. \end{aligned}$$

So F is as desired.

It remains to show that the phalanx $(\langle W, j(W)\rangle, \langle\alpha\rangle)$ is $\Omega+1$ iterable in $V[G]$. We claim that the strategy of choosing the unique cofinal wellfounded branch is winning in the length $\Omega+1$ iteration game. If not, then as in 6.14 there are properly small $\mathcal{R} \trianglelefteq W$ and $\mathcal{S} \trianglelefteq j(W)$ such that $\alpha \in \mathrm{OR}^{\mathcal{R}} \cap \mathrm{OR}^{\mathcal{S}}$, and a putative iteration tree on $(\langle\mathcal{R},\mathcal{S}\rangle, \langle\alpha\rangle)$ which is bad; that is, which has a last, illfounded model, or is of limit length but has no cofinal wellfounded branch. Since Ω is weakly compact, our bad tree has length $< \Omega$, so that its sharp exists. Using this for absoluteness purposes, as in the proof of 6.14, we can find in $V[G]$

$$\begin{aligned} \sigma : \mathcal{P} &\to \mathcal{R} \qquad (\text{where } \mathcal{R} \trianglelefteq W)\,, \\ \tau : \mathcal{Q} &\to \mathcal{S} \qquad (\text{where } \mathcal{S} \trianglelefteq j(W))\,, \end{aligned}$$

such that $\mathcal{P}$ and $\mathcal{Q}$ are of cardinality $\leq \alpha$ and

$$\sigma \restriction \alpha = \tau \restriction \alpha = \text{identity}\,,$$

together with a countable bad tree on $(\langle\mathcal{P},\mathcal{Q}\rangle, \langle\alpha\rangle)$. Now $\mathcal{P}$ is α-strong in $V[G]$, as witnessed by σ. Also, $\mathcal{Q}$ is α-strong in M, as witnessed by τ; note that $\tau \in M$ as $M^{<j(\kappa)} \subseteq M$ in $V[G]$. Since M and $V[G]$ have the same subsets of α, 6.11 and 6.14 imply that $\mathcal{Q}$ is α-strong in $V[G]$. But then the (1)$\to$(2) direction of 6.11 implies that $(\langle\mathcal{P},\mathcal{Q}\rangle, \langle\alpha\rangle)$ is $\Omega+1$ iterable, a contradiction. □

Corollary 7.2. *Let Ω be measurable, and suppose there is a pre-saturated ideal on ω_1; then $K^c \models$ There is a Woodin cardinal.*

We conjecture that the measurable cardinal is not needed in the hypotheses of 7.1 and 7.2.

B. Generic absoluteness

One of the most important consequences of the existence of large cardinals is that the truth values of sufficiently simple statements about the reals cannot be changed by forcing. For example, if there are arbitrarily large Woodin cardinals, then $L(\mathbb{R})^V \equiv L(\mathbb{R})^{V[G]}$ for all G set-generic over V. (This result is due to Hugh Woodin.) We shall show that this generic absoluteness implies that there are inner models with Woodin cardinals.

Hugh Woodin pointed out this application of §1 - §6. The key is the following lemma.

Lemma 7.3. (Woodin) *Let Ω be measurable, and suppose K^c has no Woodin cardinals. Then there is a sentence φ in the language of set theory, and a partial order $\mathbb{P} \in V_\Omega$, such that whenever G is V-generic over $\mathbb{P}$*

$$L_{\omega_1}(\mathbb{R}))^V \models \varphi \textit{ iff } (L_{\omega_1}(\mathbb{R}))^{V[G]} \not\models \varphi .$$

Proof. Here $(L_{\omega_1}(\mathbb{R}))^{V[G]} = L_{\omega_1^{V[G]}}(\mathbb{R}^{V[G]})$. Using the formula ψ described in 6.15 (2) which defines $\langle \mathcal{J}_\delta^K \mid \delta < \omega_1 \rangle$ in all generic extensions of V by posets $\mathbb{P} \in V_\Omega$, we can construct a sentence φ such that (provably in ZFC + "Ω is measurable" + "$K^c \models$ There are no Woodin cardinals") we have

$$L_{\omega_1}(\mathbb{R}) \models \varphi \textit{ iff } \omega_1 \text{ is a successor cardinal of } K .$$

Our hypotheses guarantee $(\alpha^+)^K = \alpha^+$ for some α. If $(L_{\omega_1}(\mathbb{R}))^V \not\models \varphi$, then take $\mathbb{P} = \mathrm{Col}(\omega, \alpha)$; letting G be V-generic / $\mathbb{P}$, we have $(L_{\omega_1}(\mathbb{R}))^{V[G]} \models \varphi$ by 5.18 (3). On the other hand, if $(L_{\omega_1}(\mathbb{R}))^V \models \varphi$, then take $\mathbb{P} = \mathrm{Col}(\omega, < \alpha)$ where $\alpha < \Omega$ is inaccessible; letting G be V-generic/$\mathbb{P}$, we have $(L_{\omega_1}(\mathbb{R}))^{V[G]} \not\models \varphi$. □

Theorem 7.4. (Woodin) *Suppose that Ω is measurable, and that whenever G is V-generic/$\mathbb{P}$ for some $\mathbb{P} \in V_\Omega$, $(L_{\omega_1}(\mathbb{R}))^V \equiv (L_{\omega_1}(\mathbb{R}))^{V[G]}$. Then $K^c \models$ There is a Woodin cardinal.*

It is well known that weak homogeneity can be used to obtain generic absoluteness. We can therefore use 7.4, together with standard arguments, to show

Theorem 7.5. *If every set of reals definable over $L_{\omega_1}(\mathbb{R})$ is weakly homogeneous, then letting K^c be the model constructed in* §1 *below Ω, where Ω is the least measurable cardinal, we have $K^c \models$ There is a Woodin cardinal.*

Proof. If any set is weakly homogeneous, then there is a measurable cardinal. Let Ω be the least measurable cardinal. For any weakly homogeneous tree T, let T^* be the tree for the complement coming from the Martin-Solovay construction. (The notation assumes the homogeneity measures for T are given somehow.) So

$$p[T^*] = \mathbb{R} - p[T]$$

is true in $V[G]$ whenever G is generic for $\mathbb{P}$ with card $(\mathbb{P}) <$ additivity of homogeneity measures for T, hence whenever G is generic for $\mathbb{P} \in V_\Omega$. Let

$$\begin{aligned} S_n &= \text{universal } \Sigma_n(L_{\omega_1}(\mathbb{R})) \text{ set of reals} \\ P_n &= \text{universal } \Pi_n(L_{\omega_1}(\mathbb{R})) \text{ set of reals} \end{aligned}$$

for $1 \le n < \omega$. Pick weakly homogeneous trees U_n such that $P_n = p[U_n]$ and let T_{n+1} be the canonical weakly homogeneous tree which projects to $\exists^{\mathbb{R}} p[U_n]$ in all $V[G]$

$$p[T_{n+1}] = \exists^{\mathbb{R}} p[U_n] \text{ in all } V[G]\,.$$

(Thus in V, $p[T_{n+1}] = S_{n+1}$.)

Claim. If G is $\mathbb{P}$-generic, where $\mathbb{P} \in V_\Omega$, then for all $n \ge 2$

$$V[G] \models p[U_n] = \mathbb{R} - p[T_n]\,.$$

Proof. Fix $V[G]$. Since $p[U_n] \cap p[T_n] = \emptyset$ in V, this remains true in $V[G]$ by absoluteness of wellfoundedness. On the other hand, if $x \in \mathbb{R}^{V[G]}$ and $x \notin (p[U_n] \cup p[T_n])$, then $x \in p[U_n^*] \cap p[T_n^*]$ because U_n^* and T_n^* project absolutely to the complements of the projections of U_n, T_n. But then $p[U_n^*] \cap p[T_n^*] \ne \emptyset$ in V by absoluteness of wellfoundedness. This is a contradiction as $p[U_n] = \mathbb{R} - p[T_n]$ in V. □

It follows that in all $V[G]$, G generic for $\mathbb{P} \in V_\Omega$,

$$p[U_{n+1}] = \text{universal } \Pi^1_n(A) \text{ set of reals, where } A = p[U_1]\,.$$

But now the fact that A is the universal $\Pi_1(L_{\omega_1}(\mathbb{R}))$ set of reals is a Π^1_{20} fact about A. So in all such $V[G]$

$$p[U_{n+1}] = \text{universal } \Pi_{n+1}(L_{\omega_1}(\mathbb{R})) \text{ set of reals}\,.$$

Thus for any sentence φ of the language of set theory

$$(L_{\omega_1}(\mathbb{R}))^V \models \varphi \text{ iff } (L_{\omega_1}(\mathbb{R}))^{V[G]} \models \varphi\,.$$

By Theorem 7.4, $K^c \models$ There is a Woodin cardinal, where K^c is constructed below Ω. □

Woodin has shown (unpublished) that if there is a strongly compact cardinal, then all sets of reals in $L(\mathbb{R})$ are weakly homogeneous. So we have at once:

Theorem 7.6. *Suppose there is a strongly compact cardinal, and let K^c be the model of §1 constructed below Ω, where Ω is the least measurable cardinal. Then $K^c \models$ There is a Woodin cardinal.*

We shall give a more direct proof of Theorem 7.6 in §8, a proof which does not rely on Woodin's work deriving weak homogeneity from strong compactness.

We conclude this section on generic absoluteness by re-proving a slightly weaker version of the following theorem due to Woodin:

$$\text{Con}\,(\text{ZFC} + \Delta^1_2\text{-determinacy}) \Rightarrow \text{Con}\,(\text{ZFC} + \text{There is a Woodin cardinal}).$$

Because the theory of §1 - §6 relies on the measurable cardinal cardinal Ω, we do not see how to use it to prove Woodin's theorem in full, although we believe that should be possible. We can, however, prove the theorem with its hypothesis strengthened to: Con $(ZFC + \Delta^1_2\text{-determinacy} + \forall x \in \mathbb{R}\,(x^\sharp \text{ exists}))$. Modulo the theory of §1 - §6, our proof is simpler than Woodin's.

Our proof relies on the observation that the theory of §1 - §6 uses somewhat less than a measurable cardinal. Namely, suppose A is a set of ordinals and $A^\sharp$ exists. Let c_0 be an indiscernible of $L[A]$, let $j : L[A] \to L[A]$ have critical point c_0, and let $\mathcal{U}$ be the $L[A]$-ultrafilter on c_0 given by: $X \in \mathcal{U} \Leftrightarrow c_0 \in j(X)$. Working in $L[A]$, we can construct $(K^c)^{L[A]}$ below c_0 just as we constructed K^c below Ω in §1. Let us assume that $L[A]$ satisfies: There is no proper class inner model with a Woodin cardinal. We can then conduct our proof of iterability within $L[A]$ (using 2.4 (b) rather than 2.4 (a)), and we have that indeed $(K^c)^{L[A]}$ exists and (by 2.10) is (ω, θ) iterable for all θ. Further, the proof of 1.4 shows that for $\mathcal{U}$ a.e. $\alpha < c_0$, $L[A] \models (\alpha^+)^{K^c} = \alpha^+$. (The main point here is that we don't really need $\mathcal{U} \in L[A]$ to carry out the proof; it is enough that if E_j is the $(c_0, j(c_0))$ extender over $L[A]$ derived from j, and $\mathcal{A} \in L[A]$ and $|\mathcal{A}|^{L[A]} \leq c_0$, then $E_j \cap ([j(c_0)]^{<\omega} \times \mathcal{A}) \in L[A]$. That these fragments of E_j are in $L[A]$ is well known.) This implies that $L[A] \models$ "c_0 is A_0-thick in K^c". We can therefore carry out the arguments of §3 - §6 within $L[A]$, and we get that $K^{L[A]}$ exists, is absolute for forcing over $L[A]$ with posets $\mathbb{P} \in V^{L[A]}_{c_0}$, and inductively definable over $L[A]$ as in §6. (The only serious use of the measurable cardinal Ω in these sections occurs in the proof of 4.8. Once again, it is clear from that proof that we only need the fragments $E_j \cap ([j(c_0)]^{<\omega} \times \mathcal{A})$, for $|\mathcal{A}|^{L[A]} \leq c_0$, to be in $L[A]$.) We also have that for $\mathcal{U}$ a.e. $\alpha < c_0$, $L[A] \models (\alpha^+)^K = \alpha^+$.

Theorem 7.7. (Woodin) *If $\forall x \in {}^\omega\omega$ ($x^\sharp$ exists) and all Δ^1_2 games are determined, then there is a proper class inner model with a Woodin cardinal.*

Proof. According to a theorem of Kechris and Solovay (cf. [KS]), Δ^1_2 determinacy implies that there is a real x such that for all reals $y \geq_T x$, $L[y] \models$ "All ordinal-definable games are determined". Fix such a real x, and let c_0 be the least indiscernible of $L[x]$. We may suppose that $L[x] \models$ "There is no proper class inner model with a Woodin cardinal". As we have observed, this means that $K^{L[x]}$ exists and is absolute for size $< c_0$ forcing over $L[x]$, and that for $\mathcal{U}$-a.e. $\alpha < c_0$, $L[x] \models (\alpha^+)^K = \alpha^+$, where $\mathcal{U}$ is the $L[x]$-ultrafilter on c_0 given by $x^\sharp$. Let $\alpha < c_0$ be such that $L[x] \models (\alpha^+)^K = \alpha^+$, and let

$y = \langle x, z\rangle$ where z is (a real) generic over $L[x]$ for the poset $\mathrm{Col}(\omega, \alpha)$ collapsing α to be countable. Then in $L[y]$, K exists and is inductively definable as in §6, and ω_1 is a successor cardinal of K. Moreover, OD determinacy holds in $L[y]$. Let us work in $L[y]$. Now OD determinacy implies that every OD set $A \subseteq \omega_1$ either contains or is disjoint from a club, and therefore that ω_1^V is measurable in HOD. On the other hand, $K \subseteq \mathrm{HOD}$, so since $\omega_1^V = (\alpha^+)^K$, $\omega_1^V = (\alpha^+)^{\mathrm{HOD}}$. But $\mathrm{HOD} \models \mathrm{AC}$, so $\mathrm{HOD} \models$ all measurable cardinals are inaccessible. This contradiction completes the proof. □

C. Unique branches

The Unique Branches Hypothesis, or UBH, is the assertion that if $\mathcal{T}$ is an iteration tree on V, then $\mathcal{T}$ has at most one cofinal wellfounded branch. Martin and the author showed that the negation of UBH has some logical strength, in that it implies the existence of an inner model with a Woodin cardinal and a measurable above. (Cf. [IT], §5.) Woodin then showed, in unpublished work, that if there is a nontrivial elementary $j : V_{\lambda+1} \to V_{\lambda+1}$, for some λ, then UBH fails. The gap between these two bounds on the consistency strength of -UBH is, of course, enormous. Here we shall improve the lower bound to two Woodin cardinals. (However, we must add "There is a measurable cardinal" to -UBH because the basic theory demands it.) We conjecture that -UBH is equiconsistent with the existence of two Woodin cardinals.

Theorem 7.8. *Let Ω be measurable, and suppose there is a normal iteration tree $\mathcal{T}$ on V such that $\mathcal{T} \in V_\Omega$ and $\mathcal{T}$ has distinct cofinal wellfounded branches. Then there is a proper class inner model satisfying "There are two Woodin cardinals".*

Proof. Assume toward contradiction than there is no such model.

We shall need a slight generalization of the K^c construction in §1. Let X be any transitive set, $X \in V_\Omega$ where Ω is measurable. We can form $K^c(X)$ by relativizing the construction of §1. So $\mathcal{N}_0 = X$, and all hulls used in forming $\mathfrak{C}_\omega(\mathcal{N}_\xi(X)) = \mathcal{M}_\xi(X)$ contain $X \cup \{X\}$, so that $X \in \mathcal{N}_\xi(X)$ for all ξ. We require that all extenders added to the $K^c(X)$ sequence have critical point $> \mathrm{OR} \cap X$. We require that the levels $\mathcal{N}_\xi(X)$ of the construction be "1-small above X", that is, if κ is a critical point of an extender from the $\mathcal{N}_\xi(X)$ sequence, then for no $\delta > \mathrm{OR} \cap X$ do we have $\mathcal{J}_\kappa^{\mathcal{N}_\xi(X)} \models \delta$ is Woodin. By $K^c(X)$ we mean the limit as $\xi \to \Omega$ of the $\mathcal{M}_\xi(X)$. Let us call a structure with the appropriate first order properties of the $\mathcal{M}_\xi(X)$ an X-premouse.

If there is no $\delta > (\mathrm{OR} \cap X)$ such that $K^c(X) \models \delta$ is Woodin, then as in §2 we get that $K^c(X)$ is $(\omega, \Omega+1)$ iterable "above X", that is, via extenders on its sequence and the images thereof. (All such extenders have critical point $> \mathrm{OR} \cap X$, so none of the embeddings move X.) Of course, any two $\Omega+1$ iterable-above-X X-premice have a successful coiteration. As in 1.4, we also have $(\alpha^+)^{K^c(X)} = \alpha^+$ for μ_0 a.e. $\alpha < \Omega$, where μ_0 is a normal measure on Ω. The rest of §3 - §6 adapts in an obvious way. (We shall not need §6.)

Now let $\mathcal{T}$ be our iteration tree on V having distinct cofinal wellfounded branches b and c. We have $\mathcal{T} \in V_\Omega$, where Ω is measurable. Let

$$\delta = \delta(\mathcal{T}) = \sup\{lh\ E^{\mathcal{T}}_\alpha \mid \alpha + 1 < lh\ \mathcal{T}\}.$$

By the results of §2 of [IT], whenever $f : \delta \to \delta$ and $f \in \mathcal{M}^{\mathcal{T}}_b \cap \mathcal{M}^{\mathcal{T}}_c$, then $\mathcal{M}^{\mathcal{T}}_b \models$ "δ is Woodin with respect to f". (Equivalently, $\mathcal{M}^{\mathcal{T}}_c$ satisfies this.)

Notice that $i^{\mathcal{T}}_{ob}(\Omega) = i^{\mathcal{T}}_{oc}(\Omega) = \Omega$. Working in $\mathcal{M}^{\mathcal{T}}_b$ and $\mathcal{M}^{\mathcal{T}}_c$, let us form the models

$$R = K^c(V_\delta)^{\mathcal{M}^{\mathcal{T}}_b},$$
$$S = K^c(V_\delta)^{\mathcal{M}^{\mathcal{T}}_c}.$$

Notice here that $V_\delta^{\mathcal{M}^{\mathcal{T}}_b} = V_\delta^{\mathcal{M}^{\mathcal{T}}_c}$; setting $X = V_\delta^{\mathcal{M}^{\mathcal{T}}_b}$, we have that both R and S are 1-small above X.

Claim 1. Let $\alpha > \delta$ be a successor cardinal of R such that $\mathcal{J}^R_\alpha \not\models \exists\kappa(\delta < \kappa \wedge \kappa$ is Woodin); then $\mathcal{J}^R_\alpha$ is $\Omega + 1$ iterable above X. Similarly for S.

Proof. Our "proper smallness above X" requirement on α guarantees, as in §6, that no iteration tree on $\mathcal{J}^R_\alpha$ which is above X can have distinct cofinal wellfounded branches. Our standard reflection argument (cf. 2.4 (a)) shows that it is enough to prove the following.

Subclaim. Let $\pi : \mathcal{P} \to \mathcal{J}^R_\alpha$ be elementary, with $\mathcal{P}$ countable, and let $\pi(\bar{X}) = X$. Let $\mathcal{U}$ be a countable putative iteration tree on $\mathcal{P}$; then either $\mathcal{U}$ has a last, wellfounded model, or $\mathcal{U}$ has a cofinal wellfounded branch.

Proof. Since $\mathcal{P}$ is countable, $\mathcal{P} \in \mathcal{M}^{\mathcal{T}}_b$, and of course $\mathcal{J}^R_\alpha \in \mathcal{M}^{\mathcal{T}}_b$. Since $\mathcal{M}^{\mathcal{T}}_b$ is wellfounded, an easy absoluteness argument gives us an embedding $\sigma : \mathcal{P} \to \mathcal{J}^R_\alpha$ such that $\sigma \in \mathcal{M}^{\mathcal{T}}_b$. But also $\mathcal{U} \in \mathcal{M}^{\mathcal{T}}_b$. We can therefore carry out the iterability proof of Theorem 2.5 within $\mathcal{M}^{\mathcal{T}}_b$ using the background extenders given by the construction of $R = K^c(V_\delta)^{\mathcal{M}^{\mathcal{T}}_b}$. □

Claim 2. $P(\delta) \cap R = P(\delta) \cap S$.

Proof. Let $\alpha = (\delta^+)^R$ and $\beta = (\delta^+)^S$. By Claim 1, both $\mathcal{J}^R_\alpha$ and $\mathcal{J}^S_\beta$ are $\Omega+1$ iterable above X. It follows that they have a successful coiteration above X, and since neither can move without dropping, we get $\mathcal{J}^R_\alpha \trianglelefteq \mathcal{J}^S_\beta$ or $\mathcal{J}^S_\beta \trianglelefteq \mathcal{J}^R_\alpha$. Suppose without loss of generality that $\mathcal{J}^R_\alpha \trianglelefteq \mathcal{J}^S_\beta$.

It follows that $P(\delta) \cap R \subseteq S$, so $P(\delta) \cap R \subseteq \mathcal{M}^{\mathcal{T}}_b \cap \mathcal{M}^{\mathcal{T}}_c$, so that $R \models \delta$ is Woodin. We are done if R has another Woodin cardinal above δ, so we assume otherwise. But then, whenever γ is a successor cardinal of R above δ, then $\mathcal{J}^R_\gamma \not\models \exists\kappa(\delta < \kappa \wedge \kappa$ is Woodin). Claim 1 then shows $\mathcal{J}^R_\gamma$ is $\Omega + 1$ iterable, and since this is true for all γ, R itself is $\Omega + 1$ iterable above X.

Theorem 1.4, applied within $\mathcal{M}_b^{\mathcal{T}}$ to $R = K^c(X)$, implies that $\mathcal{M}_b^{\mathcal{T}}$ satisfies "for $i_{ob}^{\mathcal{T}}(\mu_0)$ a.e. $\alpha < \Omega$, $\alpha^+ = (\alpha^+)^{K^c(X)}$". But for μ_0 a.e. $\alpha < \Omega$, $i_{ob}^{\mathcal{T}}(\alpha) = \alpha$ and $i_{ob}^{\mathcal{T}}(\alpha^+) = \alpha^+$. Thus it is true in V that for μ_0 a.e. $\alpha < \Omega$, $\alpha^+ = (\alpha^+)^R$.

Since R and $\mathcal{J}_\beta^S$ are $\Omega + 1$ iterable above X, they have a successful coiteration above X. Since R computes successor cardinals correctly μ_0 a.e., and $\mathcal{J}_\beta^S$ cannot move without dropping, $\mathcal{J}_\beta^S \trianglelefteq R$. This completes the proof of the claim. □

Inspecting the proof of claim 2, we have:

Claim 3. δ is Woodin in both R and S. Both R and S are $\Omega + 1$ iterable. Finally, $(\alpha^+)^R = (\alpha^+)^S = \alpha^+$ for μ_0 a.e. α.

Let us emphasize that R and S are $(\omega, \Omega + 1)$ iterable in V, not just in the models $\mathcal{M}_b^{\mathcal{T}}$ or $\mathcal{M}_c^{\mathcal{T}}$.

We wish to compare R with S, but first we must pass to models for which the comparison will have a large set of fixed points. Working in $\mathcal{M}_b^{\mathcal{T}}$, let R^* come from R by taking ultrapowers by the order zero total measure at each measurable cardinal of R. Thus R^* is $\mathcal{M}_b^{\mathcal{T}}$ definable (from δ and Ω), R^* is a linear iterate of R, and if $\delta < \kappa < \Omega$ and κ is strongly inaccessible in $\mathcal{M}_b^{\mathcal{T}}$, then κ is not the critical point of a total extender on the R^* sequence. Let S^* be obtained from S, working inside $\mathcal{M}_c^{\mathcal{T}}$, in a similar fashion.

Now let $(\mathcal{U}, \mathcal{V})$ be a successful coiteration of R^* with S^*, according to their unique $\Omega + 1$ iteration strategies. Since R^* and S^* compute α^+ correctly for a.e. $\alpha < \Omega$, $\mathcal{U}$ and $\mathcal{V}$ have a common last model Q. Let $j : R^* \to Q$ and $k : S^* \to Q$ be the iteration maps. Let

$$Z = \{\alpha < \Omega \mid j(i_{ob}^{\mathcal{T}}(\alpha)) = k(i_{oc}^{\mathcal{T}}(\alpha)) = \alpha\}$$

be the set of common fixed points of $j \circ i_{ob}^{\mathcal{T}}$ and $k \circ i_{oc}^{\mathcal{T}}$. We have then that $\mu_0(Z) = 1$, and for μ_0 a.e. α, Z is cofinal in α^+ and $\alpha^+ = (\alpha^+)^Q$.

Now let $\alpha_0 \in b - c$, and define

$$\begin{aligned} \beta_n &= \text{least } \gamma \in (c - \alpha_n), \\ \alpha_{n+1} &= \text{least } \gamma \in (b - \beta_n). \end{aligned}$$

Let us assume that α_0 is chosen large enough that $\delta \in \operatorname{ran} i_{\alpha_1,b}^{\mathcal{T}} \cap \operatorname{ran} i_{\beta_1,c}^{\mathcal{T}}$. It follows, of course, that $R^* \in \operatorname{ran} i_{\alpha_1,b}^{\mathcal{T}}$ and $S^* \in \operatorname{ran} i_{\beta_1,c}^{\mathcal{T}}$. Set

$$\kappa = \operatorname{crit} i_{\alpha_1,b}^{\mathcal{T}}$$

and

$$H = \text{transitive collapse of } \operatorname{Hull}^Q(V_\kappa^{\mathcal{M}_b^{\mathcal{T}}} \cup Z \cup \{\delta\}).$$

The next claim comes directly from the proof of the uniqueness theorem of §2 of [IT] (see also 6.1 of [FSIT]).

Claim 4. $\operatorname{Hull}^Q(V_\kappa^{\mathcal{M}_b^{\mathcal{T}}} \cup Z \cup \{\delta\}) \cap V_\delta^{\mathcal{M}_b^{\mathcal{T}}} = V_\kappa^{\mathcal{M}_b^{\mathcal{T}}}$.

Proof. (Sketch) For all $i \geq 1$, α_i and β_i are successor ordinals, and

$$\mathrm{crit}(E^{\mathcal{T}}_{\alpha_i - 1}) < \mathrm{crit}(E^{\mathcal{T}}_{\beta_i - 1}) < \mathrm{str}^{\mathcal{M}^{\mathcal{T}}_{\alpha_i}}(E^{\mathcal{T}}_{\alpha_i - 1})$$

and

$$\mathrm{crit}(E^{\mathcal{T}}_{\beta_i - 1}) < \mathrm{crit}(E^{\mathcal{T}}_{\alpha_{i+1} - 1}) < \mathrm{str}^{\mathcal{M}^{\mathcal{T}}_{\beta_i}}(E^{\mathcal{T}}_{\beta_i - 1}) .$$

Here $\mathrm{str}^M(E)$ is the strength of E in the model M. Now suppose t is a sequence of parameters from $V_\kappa^{\mathcal{M}^{\mathcal{T}}_b} \cup Z \cup \{\delta\}$, and

$$Q \models \exists x \in V_\delta^{\mathcal{M}^{\mathcal{T}}_b} \varphi(x, t) .$$

Let $\kappa_i = \mathrm{crit}(E^{\mathcal{T}}_{\alpha_i - 1})$ and $\nu_i = \mathrm{crit}(E^{\mathcal{T}}_{\beta_i - 1})$. Since $\sup\{\kappa_i \mid i \in \omega\} = \sup\{\nu_i \mid i \in \omega\} = \delta$, we can let i be least such that for some $x \in V_{\kappa_i}^{\mathcal{M}^{\mathcal{T}}_b}$, $Q \models \varphi[x, t]$. Fix such an x in $V_{\kappa_i}^{\mathcal{M}^{\mathcal{T}}_b}$. We claim $i = 1$; since $\kappa_1 = \kappa$ this will complete the proof of claim 4. Suppose then $i = e + 1$.

Since k is the identity on $V_\delta^{\mathcal{M}^{\mathcal{T}}_b} \cup Z \cup \{\delta\}$, we have $k(\langle x, t\rangle) = \langle x, t\rangle$, so $S^* \models \varphi[x, t]$. Let $\beta = T$-pred (β_e), and let

$$i^{\mathcal{T}}_{\beta,c}(\langle \bar{S}, \bar{t}\rangle) = \langle S^*, t\rangle .$$

Now $\nu_e = \mathrm{crit}\ i^{\mathcal{T}}_{\beta,c}$, and $\kappa_{e+1} < i^{\mathcal{T}}_{\beta,c}(\nu_e)$. Since $i^{\mathcal{T}}_{\beta,c}$ is elementary and $x \in V_{\kappa_{e+1}}^{\mathcal{M}^{\mathcal{T}}_c}$, we have $x' \in V_{\nu_e}^{\mathcal{M}^{\mathcal{T}}_c}$ such that $\bar{S} \models \varphi[x', \bar{t}]$. But then $S^* \models \varphi[x', t]$, and hence $Q \models \varphi[x', t]$.

We can now go apply the argument of the last paragraph to $i^{\mathcal{T}}_{\alpha,b}$, where $\alpha = T\text{-pred}(\alpha_e)$, using R^* and j instead of S^* and k. We get $x'' \in V_{\kappa_e}^{\mathcal{M}^{\mathcal{T}}_b}$ such that $Q \models \varphi[x'', t]$. This contradicts the minimality of i, and completes the proof of claim 4. □

Let $\pi : H \to Q$ be the collapse map, so that $\pi(\kappa) = \delta$ and $H \models \kappa$ is Woodin by claim 4. The properties of Z guarantee that $(\alpha^+)^H = \alpha^+$ for μ_0 a.e. $\alpha < \Omega$, and that in fact Ω is A-thick in H, where $A = \{\alpha < \Omega \mid \alpha \text{ is inaccessible}\}$.

Now, working in R, let

$$M = K^c(V_\kappa)^R .$$

Claim 5. $M \models \kappa$ is Woodin.

Proof. Assume otherwise; letting $\alpha = (\kappa^+)^M$, we then have that $\mathcal{J}^M_\alpha$ is properly small above $V_\kappa^{\mathcal{M}^{\mathcal{T}}_b}$. We get that $\mathcal{J}^M_\alpha$ is $\Omega + 1$ iterable (in V, not just in R) by the same argument we used to prove claim 1. But then H and $\mathcal{J}^M_\alpha$ are $V_\kappa^{\mathcal{M}^{\mathcal{T}}_b}$-premice which are $\Omega + 1$ iterable above $V_\kappa^{\mathcal{M}^{\mathcal{T}}_b}$, so they have a successful coiteration above $V_\kappa^{\mathcal{M}^{\mathcal{T}}_b}$. Since $\mathcal{J}^M_\alpha \models \kappa$ is not Woodin, there is a subset of κ which is in $\mathcal{J}_\alpha^M$ but not H. This means $\mathcal{J}_\alpha^M$ must iterate past H. On the other hand, H computes α^+ correctly for μ_0 a.e. $\alpha < \Omega$, so $\mathcal{J}_\alpha^M$ cannot iterate past H. □

Now $\kappa < \delta$, δ is Woodin in R, and $M = K^c(V_\kappa)^R$. A standard argument shows that for some ν such that $\kappa < \nu \leq \delta$, $M \models \nu$ is Woodin. (See the proof of 11.3 of [FSIT]. Thus $M \models$ There are two Woodin cardinals, and the proof of 7.8 is complete. □

D. Σ_3^1 correctness and the size of u_2

We say that a transitive model M is Σ_3^1 correct iff whenever $x \in M \cap {}^\omega\omega$ and P is a nonempty $\Pi_2^1(x)$ set of reals, then $P \cap M \neq \emptyset$. The proof of the following theorem was inspired by, and relies quite heavily upon, an idea due to G. Hjorth.

Theorem 7.9. *Suppose $K^c \models$ "There are no Woodin cardinals", and suppose there is a measurable cardinal $\mu < \Omega$; then K^c (or equivalently, K) is Σ_3^1 correct.*

The remarkable insight that there are theorems along the lines of 7.9, and the proof of the first of them, are due to Jensen (cf. [D]). Jensen's work was later extended by Mitchell ([M2]), and by Steel and Welch ([SW]). The smallness hypotheses on K in these works are, respectively: no inner model with a measurable cardinal, no inner model with a cardinal κ such that $o(\kappa) = \kappa^{++}$, and no inner model with a strong cardinal.

The smallness hypothesis on K in Theorem 7.9 is necessary. For if $K^c \models$ "There is a Woodin cardinal", then K^c is not Σ_3^1 correct. [Let $P = \{x \in {}^\omega\omega \mid x$ codes a countable, Π_2^1 -iterable premouse which is not 1-small$\}$. The existence of the measurable cardinal Ω gives $P \neq \emptyset$. On the other hand $P \cap K^c = \emptyset$, since if $\mathcal{M}$ is coded by a real in P, then $\mathcal{J}_\alpha^{K^c} \trianglelefteq \mathcal{M}$ for $\alpha = \omega_1^{K^c}$. (Cf. [PW], 3.1.)] However, if we liberalize our definition of K^c so as to allow levels which are not 1-small (but still retain some weaker smallness condition, e.g. tameness, which suffices to develop the basic theory of K^c), then we can simply drop the hypothesis that K^c satisfies "There are no Woodin cardinals" from 7.9. This is because if there are arbitrarily large $\alpha < \omega_1^{K^c}$ such that $\mathcal{J}_\alpha^{K^c}$ is not 1-small, then K^c is Σ_3^1 correct. (In fact, if x is a real coding a countable, $\omega_1 + 1$-iterable, non-1-small mouse $\mathcal{M}$ such that $y \in M$, and P is nonempty and $\Pi_2^1(y)$, then $\exists z \in P(z \leq_T x)$. This result is due to Woodin; cf. [PW], §4.)

Where we have assumed in 7.9 that there are two measurable cardinals, [D] requires only that every real has a sharp, and [M2] and [SW] require only the sharps of certain reals. We believe that it should be possible to eliminate the hypothesis that there is a measurable cardinal $< \Omega$ from 7.9. Of course, the need for Ω itself is also problematic, here and elsewhere.

Proof of 7.9. Our proof descends from a proof of Jensen's Σ^1_3 correctness theorem which is much simpler than Jensen's original proof. That simpler proof is due to Magidor.

Suppose that $K^c \models$ "There are no Woodin cardinals", and let $\mu < \Omega$ be measurable. For $\alpha \geq 1$, we let u_α be the αth uniform indiscernible relative to parameters in V_μ, that is

$$\begin{aligned} u_\alpha &= \alpha\text{th ordinal } \beta \text{ such that } \forall x \in V_\mu \\ &\quad (\beta \text{ is an indiscernible of } L[x]) \,. \end{aligned}$$

Thus $u_1 = \mu$. Magidor's argument is based on the following lemma.

Lemma 7.10. (Magidor) *Suppose $u^K_2 = u_2$; then there is a tree $T_2 \in K$ such that $p[T_2]$ is the universal Π^1_2 set of reals, and thus K is Σ^1_3 correct.*

Proof. (Sketch) We first show that for all α, $u^K_\alpha = u_\alpha$. The proof is by induction on α; the cases $\alpha = 1$ and α is a limit are trivial. Let $\alpha = \beta + 1$. Let

$$n(\gamma, x) = \text{least indiscernible of } L[x] \text{ which is } > \gamma \,.$$

We have

$$u_{\beta+1} = \sup\{n(u_\beta, x) \mid x \in V_\mu\} \,,$$

and

$$\begin{aligned} u_2 &= \sup\{n(u_1, x) \mid x \in V_\mu\} \\ &= \sup\{n(u_1, x) \mid x \in V^K_\mu\} \,, \end{aligned}$$

since $u_2 = u^K_2$. But then for any $x \in V_\mu$, we can find $y \in V^K_\mu$ so that $n(u_1, x) < n(u_1, y)$, and thus $n(u_\beta, x) < n(u_\beta, y)$ by the uniform indiscernibility of the u_η's. It follows that

$$u_{\beta+1} = \sup\{n(u_\beta, x) \mid x \in V^K_\mu\} \,,$$

as desired.

It is well known that for any ordinal η, these are an $x \in V_\mu$ and a term τ and uniform indiscernibles $u_{\alpha_0} < \cdots < u_{\alpha_n} \leq \eta$ such that $\eta = \tau^{L[x]}(u_{\alpha_0} \cdots u_{\alpha_n})$. (This result is due to Solovay; the proof is an easy induction on η.) Since $u_\alpha = u^K_\alpha$ for all α, we can take $x \in K$ in the above.

By T_2, we mean the Martin-Solovay tree for Π^1_2 constructed as follows. Let $\mathbf{L} = \bigcup\{L[x] \mid x \in V_\mu\}$. Let S on $\omega \times \omega \times \mu$ be the Shoenfield tree for a Buniversal Σ^1_2 set. For u, $v \in \omega^{<\omega}$ such that $\text{dom}(u) = \text{dom}(v)$, let $S_{(u,v)} = \{w \mid (u, v, w) \in S\}$. We define an ultrafilter on $P(S_{(u,v)}) \cap \mathbf{L}$ as follows. For $X \subseteq \mu$ and $n < \omega$, let $[X]^n = \{\langle \alpha_0 \cdots \alpha_{n-1} \rangle \mid \alpha_0 < \alpha_1 < \cdots < \alpha_{n-1} \wedge \forall i < n(\alpha_i \in X)\}$. Letting $n = \text{dom}(u) = \text{dom}(v)$, there is a unique permutation $(i_0, \ldots, i_{n-1})$ of n such that $S_{(u,v)} = \{\langle \alpha_{i_0} \cdots \alpha_{i_{n-1}} \rangle \mid \langle \alpha_0 \cdots \alpha_{n-1} \rangle \in [\mu]^n\}$. For $A \subseteq S_{(u,v)}$ with $A \in \mathbf{L}$, we put

$$\mu_{(u,v)}(A) = 1 \Leftrightarrow \quad \exists C(C \text{ is club in } \mu \wedge \forall \langle \alpha_0 \cdots \alpha_{n-1} \rangle \in [C]^n \\ (\langle \alpha_{i_0} \cdots \alpha_{i_{n-1}} \rangle \in A)).$$

Since $\forall x \in V_\mu(x^\sharp \text{ exists})$, $\mu_{(u,v)}$ is an ultrafilter on $P(S_{(u,v)}) \cap \mathbf{L}$. If $u \subseteq r$ and $v \subseteq s$, then $\mu_{(u,v)}$ is compatible with $\mu_{(r,s)}$, so we have a natural embedding

$$\pi_{(u,v),(r,s)} : \mathrm{Ult}(\mathbf{L}, \mu_{(u,v)}) \to \mathrm{Ult}(\mathbf{L}, \mu_{(r,s)}).$$

The ultrapowers here are formed using functions in $\mathbf{L}$. The result of Solovay mentioned above yields

$$\langle u_{i_0+1}, \ldots, u_{i_{n-1}+1} \rangle = [\text{identity}]_{\mu_{u,v)}},$$

where $(i_0, \ldots, i_{n-1})$ is the permutation of $n = \mathrm{dom}(u)$ used to define $\mu_{(u,v)}$, and the u_i's are the uniform indiscernibles. By convention, $\mu_{(\emptyset,\emptyset)}$ is principal and $\mathrm{Ult}(\mathbf{L}, \mu_{(\emptyset,\emptyset)}) = \mathbf{L}$. We then have:

$$\mu_{(u,v)}(A) = 1 \quad \text{iff} \quad \langle u_{i_0+1}, \ldots, u_{i_{n-1}+1} \rangle \in \pi_{(\emptyset,\emptyset),(u,v)}(A).$$

Except for the fact that they are not total on V, the measures $\mu_{(u,v)}$ witness the weak homogeneity of S. In particular $x \in p[S]$ iff $\exists y \in {}^\omega\omega$ (the direct limit of the $\mathrm{Ult}(\mathbf{L}, \mu_{(x \restriction n, y \restriction n)})$ under the $\pi_{(x\restriction n, y \restriction n),(x \restriction n+1, y \restriction n+1)}$ is wellfounded). The tree T_2 builds a real x on one coordinate, and proves $x \notin p[S]$ on the other by showing continuously that all associated direct limits are illfounded. More precisely, let $\langle r_i \mid i \in \omega \rangle$ enumerate $\omega^{<\omega}$ so that $r_0 = \emptyset$ and $r_i \subseteq r_j \Rightarrow i \leq j$, and put for $u \in \omega^{<\omega}$ with $\mathrm{dom}(u) = n$,

$$(u, \langle \alpha_0, \ldots, \alpha_{n-1} \rangle) \in T_2 \text{ iff } \alpha_0 = \mu \wedge \forall i < j \leq n-1 \\ (r_i \subsetneq r_j \Rightarrow \pi_{(u \restriction \mathrm{dom}\, r_i, r_i)(u \restriction \mathrm{dom}\, r_j, r_j)}(\alpha_i) > \alpha_j).$$

Then $p[T_2] = {}^\omega\omega - p[S]$.

Since $K \models \forall x \in V_\mu(x^\sharp \text{ exists})$, we can form T_2^K inside K. In order to see that $T_2^K = T_2$, we must see that for any $u, v \in \omega^{<\omega}$ with $\mathrm{dom}(u) = \mathrm{dom}\, v$

$$\pi^K_{(\emptyset,\emptyset),(u,v)}(\mu) = \pi_{(\phi,\phi),(u,v)}(\mu),$$

and if $u \subseteq r$ and $v \subseteq s$ and $\mathrm{dom}(r) = \mathrm{dom}\, s$,

$$\pi^K_{(u,v),(r,s)} \restriction \mu^* = \pi_{(u,v),(r,s)} \restriction \mu^*$$

for $\mu^* = \pi_{(\emptyset,\emptyset),(r,s)}(\mu)$. Now clearly, $\mu^K_{(u,v)} = \mu_{(u,v)} \cap K$ for all (u, v). We are done, then, if we show that for any (u, v) and $f : S_{(u,v)} \to \mu$ such that $f \in L$, $[f]_{\mu_{(u,v)}}$ has a representative in K. We may assume that for some $x \in V_\mu$ and term τ, $f(w) = \tau^{L[x]}[w]$ for all $w \in S_{(u,v)}$. Let $w^* = [\text{identity}]_{\mu_{(u,v)}} < K_{U_{n+1}}$, where $n = \mathrm{dom}(v)$, so by the result of Solovay mentioned above, applied inside K, we can find a $y \in V^K_\mu$ and a term σ such that $\sigma^{L[y]}[w^*] = \tau^{L[x]}[w^*]$. It follows that for $\mu_{(u,v)}$ a.e. w, $\sigma^{L[y]}[w] = \tau^{L[x]}[w]$. Letting $g(w) = \sigma^{L[y]}[w]$ for all $w \in S_{(u,v)}$, we have $g \in K$ and $[g]_{\mu_{(u,v)}} = [f]_{\mu_{(u,v)}}$.

This completes the proof of 7.10 □

Now let

$$\mathcal{F} = \{\mathcal{M} \in V_\mu \mid \mathcal{M} \text{ is } \Omega + 1 \text{ iterable and properly small}\}.$$

Recall from 6.12 that a premouse is properly small just in case it satisfies "There are no Woodin cardinals" and "There is a largest cardinal". There can be at most one cofinal wellfounded branch in an iteration tree based on a properly small premouse, so any $\mathcal{M} \in \mathcal{F}$ has a unique $\Omega + 1$ iteration Bstrategy $\Sigma_{\mathcal{M}}$. For $\mathcal{M}$, $\mathcal{N} \in \mathcal{F}$, let $\mathcal{P}$ and $\mathcal{Q}$ be the last models of $\mathcal{T}$ and $\mathcal{U}$, where $(\mathcal{T}, \mathcal{U})$ is the unique successful $(\Sigma_{\mathcal{M}}, \Sigma_{\mathcal{N}})$ coiteration of $\mathcal{M}$ with $\mathcal{N}$. We define $\mathcal{M} \leq^* \mathcal{N}$ iff $\mathcal{P} \trianglelefteq \mathcal{Q}$. Thus $\leq^*$ is just the usual mouse order, restricted to $\mathcal{F}$. The Dodd-Jensen lemma implies that $\leq^*$ is a prewellorder. Set

$$\delta = \text{order type of } (\mathcal{F}, \leq^*).$$

Also, for $\mathcal{M} \in \mathcal{F}$, let $|\mathcal{M}|_{\leq^*}$ be the rank of $\mathcal{M}$ in the prewellorder $\leq^*$.

The following lemma is part of the folklore.

Lemma 7.11. $\delta \leq u_2^K$.

Proof. It is easy to see that if U is any normal ultrafilter on μ, then $(\alpha^+)^K = \alpha^+$ for U a.e. $\alpha < \mu$. B(We prove this as part of the proof of lemma 8.15 in the next section.) It follows that $K \cap \mathcal{F}$ is $\leq^*$-cofinal in $\mathcal{F}$. For let $\mathcal{M} \in \mathcal{F}$, and let $(\mathcal{T}, \mathcal{U})$ be the successful coiteration of $\mathcal{M}$ with $\mathcal{J}_\mu^K$ determined by $\Omega + 1$ iteration strategies for the two mice. Since $\mathcal{J}_\mu^K$ computes successor cardinals correctly almost everywhere, $\max(lh\, \mathcal{T}, lh\, \mathcal{U}) < \mu$, and the last model $\mathcal{P}$ of $\mathcal{T}$ is an initial segment of the last model of $\mathcal{U}$. Let $\alpha < \mu$ be a successor cardinal of K and such that $lh\ E_\xi^{\mathcal{U}} < \alpha$ for all $\xi + 1 < lh\, \mathcal{U}$; then we can regard $\mathcal{U}$ as tree on $\mathcal{J}_\alpha^K$, so that $(\mathcal{T}, \mathcal{U})$ demonstrates that $\mathcal{M} \leq^* \mathcal{J}_\alpha^K$.

It suffices then to show that if $\mathcal{M} \in K \cap \mathcal{F}$, then $|\mathcal{M}|_{\leq^*} < u_2^K$. Fix $\mathcal{M}$, and let G be V-generic for $\mathrm{Col}(\omega, < \mu)$, and let x_0 be a real coding $\mathcal{M}$ in $V[G]$. Choose x_0 to be generic over $L[\mathcal{M}]$, so that $(\mu^+)^{L[x_0]} = (\mu^+)^{L[\mathcal{M}]} < u_2^K$. For x and y reals in $V[G]$, let

$R(x, y)$ iff (x and y code properly small premice$\mathcal{M}_x$ *and* $\mathcal{M}_y$, and there is a successful coiteration $(\mathcal{T}, \mathcal{U})$ of $\mathcal{M}_x$with $\mathcal{M}_y$ such that $\mathcal{T}$ and $\mathcal{U}$ are simple, and the last model of $\mathcal{T}$ is a proper initial segment of that of $\mathcal{U}$),

and let

$$S(x, y) \text{ iff } R(x, y) \cap R(y, x_0).$$

It is easy to check that S is a $\Sigma_2^1(x_0)$ relation on the reals in $V[G]$.

Claim. S is wellfounded.

Proof. Suppose not, and let $S = \dot{S}_G$, where $\dot{S}$ represents the natural definition of S from $\mathcal{M}$ over $V[G]$. Working in V, we can construct a countable, transitive P and an elementary $\pi : P \to V_\Omega$ with $\pi(\bar{\mathcal{M}}) = \mathcal{M}$ and $\pi(\bar{\mu}) = \mu$ for some $\bar{\mathcal{M}}$, $\bar{\mu}$. Let $\bar{G}$ in V be P-generic for $\mathrm{Col}(\omega, < \bar{\mu})$. Then $\dot{S}_{\bar{G}}$ is illfounded, and this implies that the mouse order below $\mathcal{M}$ is illfounded. Since $\pi \restriction \bar{\mathcal{M}} : \bar{\mathcal{M}} \to M$, and $\mathcal{M} \in \mathcal{F}$, this is a contradiction.

Since S is wellfounded and $\Sigma^1_2(x_0)$, its rank is $< (\mu^+)^{L[x_0]}$ by the Kunen-Martin theorem. Clearly, $(|\mathcal{M}|_{\leq^*})^V$ is less than or equal to the rank of S. This proves 7.11. □

In view of 7.10 and 7.11, we would like to show that $\delta = u_2$. The key idea for doing this is due to Greg Hjorth.

Lemma 7.13. (Hjorth) *Suppose $\delta < u_2$; then there is a set $M \in V_\mu$ such that $\mathcal{F} \cap L[M]$ is $\leq^*$-cofinal in $\mathcal{F}$.*

Proof. Since $\delta < u_2$, we have an $x \in V_\mu$ and a term τ such that $\delta = \tau^{L[x]}[x, \mu]$. Now let $x \in V_\eta$ where $\eta < \mu$, and let $(Z, \epsilon) < (V_\Omega, \epsilon)$ be such that $\mathrm{card}(Z) < \mu$, $V_\eta \subseteq Z$, and $\mu \in Z$. Let M be the transitive collapse of Z, $\pi : M \to V_\Omega$ the collapse map, and $\pi(\bar{\mu}) = \mu$. Let $\bar{U}$ be such that $M \models \bar{U}$ is a normal ultrafilter on $\bar{\mu}$, let N be the μth (linear) iterate of M by $\bar{U}$ and its images, and let $i : M \to N$ be the iteration map. By an argument due to Jensen, there is an embedding $\sigma : N \to V_\Omega$ such that $\pi = \sigma \circ i$. Now $i(\bar{\mu}) = \mu$, so $\sigma(\mu) = \sigma(i(\bar{\mu})) = \pi(\bar{\mu}) = \mu$. We also have $\pi \restriction V_\eta = i \restriction V_\eta =$ identity, so $\sigma \restriction V_\eta =$ identity, so $\sigma(x) = x$. Thus $\sigma(\tau^{L[x]}[x, \mu]) = \tau^{L[x]}[x, \mu]$; that is, $\sigma(\delta) = \delta$. It follows that $(\mathcal{F}^N, \leq^{*\,N})$ has order type δ. Now if $\mathcal{P} \in \mathcal{F}^N$, then $\sigma \restriction \mathcal{P} : \mathcal{P} : \mathcal{P} \to \sigma(\mathcal{P})$ and $\sigma(\mathcal{P}) \in \mathcal{F}$, and thus $\mathcal{P}$ is $\Omega + 1$-iterable. Thus $\mathcal{F}^N \subseteq \mathcal{F}$. It is easy to see that $(\leq^*)^N = \leq^* \cap N$. Since $N \in L[M]$, $\mathcal{F}^N \subseteq \mathcal{F} \cap L[M]$, and we are done. □

We will actually use the proof of 7.12, rather than the lemma itself.

So far we haven't worked with K above μ, and indeed Hjorth formulated his lemma with $\mu = \Omega$. But now let M and N be as in the proof of 7.12. We would be done if we could find $\mathcal{P} \in \mathcal{F}$ such that $\forall Q \in \mathcal{F}^N (Q \leq^* \mathcal{P})$. There is a natural candidate for such a $\mathcal{P}$, namely K^M. (K^M is not actually properly small, but this problem is easily finessed.) Of course, the iteration map $i : K^M \to K^N$ comes from an "external" iteration of all of M, but suppose we could absorb its action into an internal iteration of K^M. We'd be done. Since $\mathrm{crit}(i) = \bar{\mu}$, we must use the part of K^M above $\bar{\mu}$ to do this. So we must work with $\mu < \Omega$.

The following lemma is the key to absorbing the map from K^M to K^N into an iteration of K^M. Its proof borrows Lemma 8.2 from §8, a lemma we originally proved as part of the proof of 7.9.

Lemma 7.13. *Let $j : V \to \mathit{Ult}(V, U)$, where U is a normal ultrafilter on μ; then there are almost normal iteration trees $\mathcal{T}$ on K and $\mathcal{U}$ on $j(K)$,* *having common last model Q and associated embeddings $k : K \to Q$ and $\ell : j(K) \to Q$, such that $k = \ell \circ j$.*

Proof. Let W be a weasel such that Ω is thick in W, and W has the hull property at all $\alpha < \Omega$. Lemma 4.5 shows that such a weasel exists. By Lemma 8.2, there is an iteration tree $\mathcal{T}_0$ on K having last model W whose associated embedding $t_0 : K \to W$ satisfies BDef$(W) = t_0''K$. In fact, $\mathcal{T}_0$ is a linear iteration by normal measures. Notice that $j(\mathcal{T}_0)$ is an iteration tree on $j(K)$ with last model $j(W)$ and associated embedding $j(t_0)$. Since the class of fixed points of j is thick in W, Ω is thick in $j(W)$ and Def$(j(W)) = j''$ Def(W).

Now let $(\mathcal{T}_1, \mathcal{U}_1)$ be the successful coiteration of W with $j(W)$, using their unique $\Omega + 1$ iteration strategies, and let Q be the common last model of $\mathcal{T}_1$ and $\mathcal{U}_1$. BLet $t_1 : W \to Q$ and $u : j(W) \to Q$ be the associated iteration maps.

We have the diagram:

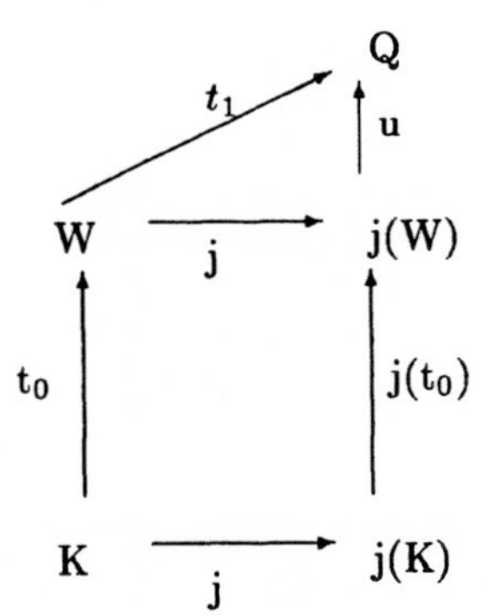

The bottom rectangle commutes: $j \circ t_0 = j(t_0) \circ j$ because j is elementary on V. The upper "triangle" may not commute, but it commutes on ran(t_0), since

$$\begin{aligned} t_1''(t_0''K) &= t_1'' \operatorname{Def}(W) = \operatorname{Def}(Q) \\ &= u'' \operatorname{Def}(j(W)) = u''(j'' \operatorname{Def}(W)) \\ &= u''(j''(t_0''K)). \end{aligned}$$

It follows that, setting $\mathcal{T} = \mathcal{T}_0{}^\frown \mathcal{T}_1$, $k = t_1 \circ t_0$, $\mathcal{U} = j(\mathcal{T}_0){}^\frown \mathcal{U}_1$, and $\ell = u \circ j(t_0)$, the conclusion of 7.13 holds. □

We can now complete the proof of Theorem 7.9. By 7.10 and 7.11 it is enough to show $\delta = u_2$, so assume $\delta < u_2$. Let M, N, i, and $\bar{\mu}$ be as in the proof of 7.12. So $M \in V_\mu$, and $i : M \to N$ is the iteration map coming from hitting a normal meaAsure of M on $\bar{\mu}$ repeatedly, μ times in all. LeAt $i_{\alpha\beta} : M_\alpha \to M_\beta$ be the natural map, where M_α and M_β are the αth and βth iterates of M. So $i = i_{0\mu}$.

We define premice Q_α, for $\alpha \leq \mu$, by induction on α. We shall have that $Q_{\alpha+1}$ is the last model on an almost normal iteration tree $\mathcal{T}_\alpha$ on Q_α, with an associated iteration map $k_{\alpha,\alpha+1} : Q_\alpha \to Q_{\alpha+1}$. We shall simultaneously define embeddings $\ell_\alpha : K^{M_\alpha} \to Q_\alpha$ so that for $\alpha \leq \beta \leq \mu$

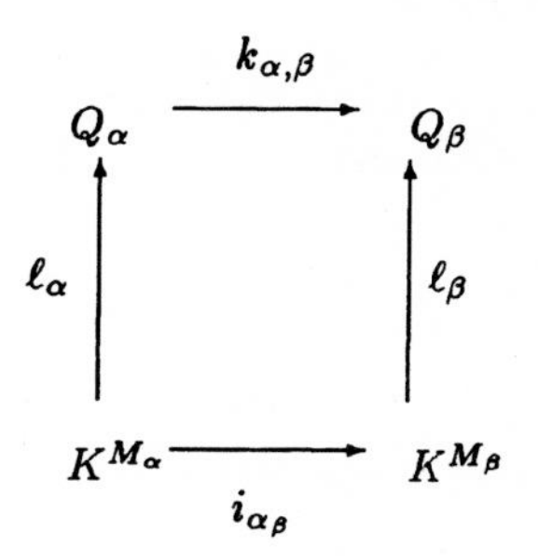

commutes. (Here we are setting $k_{\alpha,\gamma+1} = k_{\gamma,\gamma+1} \circ k_{\alpha\gamma}$, and $k_{\alpha\lambda} : Q_\alpha \to Q_\lambda$ to be the canonical embedding into $Q_\lambda = \text{dir lim}_{\alpha<\lambda} Q_\alpha$ for λ limit.)

Set $Q_0 = K^{M_0}$ and $\ell =$ identity. Now, given Q_α and ℓ_α, we apply 7.13 inside the model M_α to the ultrapower which produces $M_{\alpha+1}$. This gives an almost normal iteration tree $\mathcal{T}$ on K^{M_α} with last model Q and iteration map $k : K^{M_\alpha} \to Q$, and an embedding $\ell : K^{M_{\alpha+1}} \to Q$ such that $k = \ell \circ i_{\alpha,\alpha+1}$. Note $\mathcal{T} \in M_\alpha$. Let $\mathcal{T}_\alpha = \ell_\alpha \mathcal{T}$ be the result of copying $\mathcal{T}$ to a tree on Q_α, and let $Q_{\alpha+1}$ be the last model of $\mathcal{T}_\alpha$. (K^{M_α} is a model of ZFC, $\mathcal{T}$ doesn't drop on its main branch, and k and ℓ are fully elementary. So, by induction, all Q_γ are ZFC models, no $\mathcal{T}_\gamma$ drops on its main branch, and all $k_{\eta\gamma}$ and ℓ_γ are fully elementary. So we can copy.) Let $u : Q \to Q_{\alpha+1}$ be given by the copy construction, and $\ell_{\alpha+1} = u \circ \ell$. The commutative diagram below summarizes the construction of $Q_{\alpha+1}$ and $\ell_{\alpha+1}$:

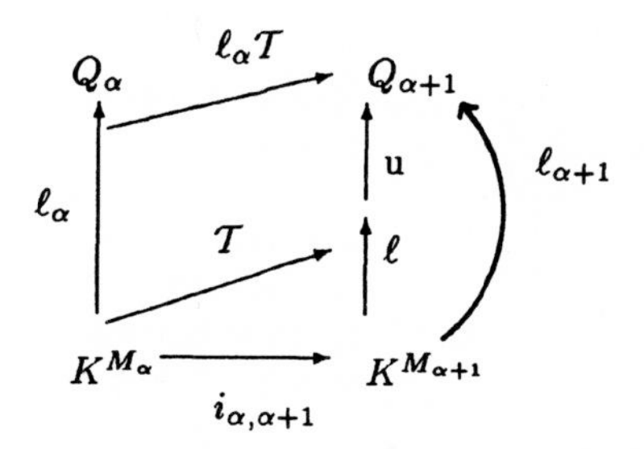

For λ a limit $\leq \mu$, let $\ell_\lambda(i_{\alpha\lambda}(x)) = k_{\alpha\lambda}(\ell_\alpha(x))$ whenever $\alpha < \lambda$ and $x \in K^{M_\alpha}$. This completes the inductive definition of Q_α and ℓ_α.

The Q_α's are not properly small, but we can easily finesse this problem. Let $\psi : V \to \text{Ult}(V, U)$ be the canonical embedding, where U is a normal ultrafilter on Ω. Let $\mathcal{P} = \mathcal{J}_\alpha^{\psi(K)}$, where $\alpha = \Omega^+ = (\Omega^+)^{\psi(K)}$. Clearly, $\mathcal{P}$ is properly small, its largest cardinal being Ω, and $K = \mathcal{J}_\Omega^{\mathcal{P}}$. Also, $\mathcal{P}$ is $\Omega+1$ iterable in V, since $\mathcal{P}$ is $\psi(\Omega+1)$ iterable in $\mathrm{Ult}(V,U)$ and $\mathrm{Ult}(V,U)$ is closed under Ω-sequences. It follows that any iteration tree on K of length $\leq \Omega+1$ which is built according to the unique $\Omega+1$ iteration strategy for K can be regarded as an iteration tree on $\mathcal{P}$. Now we can assume that the hull $Z \prec V_\Omega$ collapsing to M is such that $Z = Y \cap V_\Omega$ for some $Y \prec V_\theta$ (for $\theta > \Omega$ large) with $\Omega, \mathcal{P} \in Y$. Let M' be the transitive collapse of Y and Q_0' be the image of $\mathcal{P}$ under this collapse. Thus $Q_0' \in \mathcal{F}$ and $Q_0 = \mathcal{J}_\alpha^{Q_0'}$, where α is the collapse of Ω. We can interpret $\mathcal{T}_0$ as a tree on Q_0' according to its unique $\Omega+1$ iteration strategy, and let Q_1' be the last model of $\mathcal{T}_0$, so interpreted. Then $Q_1 = \mathcal{J}_\alpha^{Q_1'}$, where α is the largest cardinal of Q_1', and $Q_1' \in \mathcal{F}$. Proceeding similarly by induction, we define Q_α' for $\alpha \leq \mu$ so that $Q_\alpha = \mathcal{J}_\beta^{Q_\alpha'}$ for β the largest cardinal of Q_α'.

Now let $\mathcal{R} \in \mathcal{F}^N$. Working in N, we see that $\mathcal{R} \leq^* \mathcal{J}_\beta^{K^N}$ for some $\beta < \mu$. Since K^N is elementarily embedded into Q_μ by ℓ_μ, and Q_μ' is an almost normal iterate of Q_0' by its unique $\Omega+1$ iteration strategy, $\mathcal{R} \leq^* Q_0'$. Thus $\mathcal{F}^N$ is not $\leq^*$-cofinal in $\mathcal{F}$; Q_0' is an upper bound. The argument in the proof of 7.12 now yields a contradiction. □

We can use our Σ_3^1 correctness theorem to show that certain apparently weak consequences of $\boldsymbol{\Delta}_2^1$ determinacy actually imply Δ_2^1 determinacy. The ideas here are due to A. S. Kechris; what we have contributed is just Theorem 7.9.

Corollary 7.14. *Suppose $\forall x \in {}^\omega\omega$ ($x^\sharp$ exists), and $\forall x \in {}^\omega\omega$ (the class of $\Sigma_3^1(x)$ subsets of ω has the separation property). Then $\boldsymbol{\Delta}_2^1$ determinacy holds.*

Proof. We show that Δ_2^1 determinacy holds; the proof relativizes routinely to an arbitrary real. By a theorem of Woodin, it is enough to show that there is a transitive proper class model M and an ordinal δ such that $M \models \delta$ is Woodin, and $V_{\delta+1}^M$ is countable.

Let x be a real which codes up witnesses to all true Σ_3^1 sentences; that is, let x be such that whenever P is a nonempty Σ_3^1 set of reals, then $\exists y \in P$ $(y \leq_T x)$. Using the Jensen-Mitchell Σ_3^1 correctness theorem, we get a proper class model N such that $x \in N$ and $N \models$ "There is are two measurable cardinals". For if there is no such N, then $K_{DJ}(x)$ is Σ_3^1 correct, where $K_{DJ}(x)$ is the Dodd-Jensen-Mitchell core model for two measurable cardinals, relativised to x. Now $K_{DJ}(x) \models$" There is a $\Delta_3^1(x)$-good wellorder of $\mathbb{R}$", and thus $K_{DJ}(x) \models$ "There are $\Sigma_3^1(x)$ sets A, $B \subseteq \omega$ such that $A \cap B = \emptyset$ and for all $\Delta_3^1(x)$ sets C, $A \subseteq C \Rightarrow B \cap C \neq \emptyset$". Since $K_{DJ}(x)$ is Σ_3^1 correct, there really are such sets A and B, and thus $\Sigma_3^1(x)$ separation fails.

Now let N be as described in the previous paragraph, and let $N \models$ "μ and Ω are measurable", where $\mu < \Omega$. If $(K^c)^N \models$" there is a Woodin cardinal", then we get the desired proper class model M with one Woodin cardinal δ such that $V^M_{\delta+1}$ is countable. (Let P be the transitive collapse of a countable elementary submodel of $V_{\Omega+\omega}$, and $i : P \to P_\infty$ the result of iterating a normal measure on the image under collapse of Ω through OR, and let $M = i((K^c)^P)$.) But if $(K^c)^N$ satisfies that there are no Woodin cardinals, then K^N is Σ^1_3 correct in N by 7.9. The choice of x guarantees that, since $\Sigma^1_3 \cap P(\omega)$ has the separation property in V, it has the separation property in N. The correctness of K^N implies that $\Sigma^1_3 \cap P(\omega)$ has the separation property in K^N. But $K^N \models$ "$\mathbb{R}$ has a Δ^1_3-good wellorder", so $K^N \models$ "$\Sigma^1_3 \cap P(\omega)$ does not have the separation property". □

If $\Pi^1_3 \cap P({}^\omega\omega)$ has the reduction property, then for all $x \in {}^\omega\omega$ of sufficiently large Turing degree, $\Sigma^1_3(x) \cap P(\omega)$ has the separation property. [Let (A, B) reduce a universal pair of Π^1_3 subsets of ${}^\omega\omega \times {}^\omega\omega$. Then whenever A and B are $\Pi^1_3(x)$, $\Sigma^1_3(x) \cap P(\omega)$ has the separation property.] Thus the proof of 7.14 shows that $\forall x \in {}^\omega\omega$ ($x^\sharp$ exists) + "$\Pi^1_3 \cap P({}^\omega\omega)$ has the reduction property" implies Δ^1_2 determinacy. We do not know whether $\forall x \in {}^\omega\omega$ ($x^\sharp$ exists) + "$\Sigma^1_3 \cap P({}^\omega\omega)$ has the separation property" implies Δ^1_2 determinacy.

We conjecture that $\forall x \in {}^\omega\omega$ ($x^\sharp$ exists) plus "$\Sigma^1_3 \cap P(\omega)$ has the separation property" implies Δ^1_2 determinacy. If one tries to prove this lightface refinement of 7.14 by the method of 7.14, then the fact that our Σ^1_3 correctness theorem required two measurable cardinals, (rather than none) becomes a problem.

Another application of our Σ^1_3 correctness theorem in "reverse descriptive set theory" can be found in [Hj], where Hjorth uses it to show that $\boldsymbol{\Pi}^1_2$ Wadge determinacy implies $\boldsymbol{\Pi}^1_2$ determinacy.

A problem which is closely related to the Σ^1_3 correctness problem is: what is the consistency strength of ZFC + $\forall x \in {}^\omega\omega(x^\sharp$exists) + $\boldsymbol{\delta}^1_2 = \omega_2$? Woodin has shown that the strength of ZFC+ "there is a Woodin cardinal with a measurable cardinal above it" is an upper bound. It is shown in [SW] that ZFC + "There is a strong cardinal" is a lower bound. We conjecture that the lower bound can be improved to ZFC + "There is a Woodin cardinal". Unfortunately, our proof of 7.9 does not seem to help with this conjecture, because of our use of the measurable cardinals μ and Ω. One wants to replace μ with ω_1 (and V_μ with HC), and avoid Ω altogether, and we don't see how to do this. However, our proof of 7.9 does give the consistency strength lower bound ZFC + "There is a Woodin cardinal" for a certain variant of ZFC + "$\forall x \in {}^\omega\omega(x^\sharp$ exists) + $\boldsymbol{\delta}^1_2 = \omega_2$" which we now explain.

Let $\mu < \Omega$ be measurable, and let u_α be the αth uniform indiscernible relative to elements of V_μ, as in the proof of 7.9. Notice that in $V^{\mathrm{Col}(\omega,<\mu)}$, μ_α is the αth uniform indiscernible relative to reals, and so $u_2 = (\boldsymbol{\delta}^1_2)^{V^{\mathrm{Col}(\omega,<\mu)}}$. One can ask whether $V^{\mathrm{Col}(\omega,<\mu)} \models \boldsymbol{\delta}^1_2 = \omega_2$; we do not know whether it is consistent relative to any large cardinal hypothesis that this be true. But if we replace $V^{\mathrm{Col}(\omega,<\mu)}$ by its $L(\mathbb{R})$, then the resulting proposition follows from $AD^{L(\mathbb{R})}$ in $V^{\mathrm{Col}(\omega,<\mu)}$, which of course holds if there are enough Woodin cardinals in V. We now show that "$V^{\mathrm{Col}(\omega,<\mu)} \models (L(\mathbb{R}) \models \delta^1_2 = \omega_2)$" is at least as strong as the existence of one Woodin cardinal.

Theorem 7.15. *Let $\mu < \Omega$ be measurable, and suppose $V^{\mathrm{Col}(\omega,<\mu)} \models \delta^1_2 = \omega_2^{L(\mathbb{R})}$; then $K^c \models$ There is a Woodin cardinal.*

Proof. Suppose $K^c \models$ There are no Woodin cardinals. Letting u_α be the αth uniform indiscernible relative to parameters in V_μ, we have $u_2^K = u_2^V$ from the proof of 7.9. Now let G be V-generic/$\mathrm{Col}(\omega, < \mu)$. It is easy to see that u_2^V is the second uniform indiscernible relative to reals in $V[G]$, so that $u_2^V = (\delta^1_2)^{V[G]}$. Thus $u_2^K = \omega_2^{L(\mathbb{R}^*)}$, where $\mathbb{R}^* = \mathbb{R}^{V[G]}$. On the other hand, $\mathcal{J}^K_\mu$ is $\Sigma_\omega(L_\mu(\mathbb{R}^*))$ definable, by §6 and the fact that $K = K^{V[G]}$. Since $\mathcal{J}^K_\mu$ is essentially a subset of μ, we get $u_2^K < \omega_2^{L(\mathbb{R}^*)}$, a contradiction. □

§8. Embeddings of K

In this section we prove some general theorems concerning embeddings of K. We also use these theorems and the ideas behind them to give another proof that if there is a strongly compact cardinal, then there is an inner model with a Woodin cardinal.

In his work on the core model for sequences of measures ([M1] , [M ?]), Mitchell has shown that if there is no inner model satisfying $\exists\kappa(o(\kappa) = \kappa^{++})$, then for any universal weasel M there is an elementary $j : K \to M$; moreover, for any weasel M and elementary $j : K \to M$, j is the iteration map associated to some (linear) iteration of K. Thus the class of embeddings of K is precisely the class of iteration maps, and the class of range models for such embeddings is precisely the class of universal weasels. It follows at once that if there is no inner model satisfying $\exists\kappa(o(\kappa) = \kappa^{++})$, then any $j : K \to K$ is the identity; that is, K is "rigid".

Mitchell's results extend the original Dodd-Jensen theorem ([DJ1]) that if there is no inner model with a measurable cardinal, then whenever $j : K \to M$ is elementary, $M = K$ and $j =$ identity. The authors of [DJKM] strengthen Mitchell's results by weakening their non-large-cardinal hypothesis to "There is no inner model with a strong cardinal". We shall also prove such a strengthening of Mitchell's results, in Theorem 8.13 below.

The situation becomes more complicated once one gets past strong cardinals. We shall see that it is consistent with "There is no inner model having two strong cardinals" that there is a universal weasel which is not an iterate of K, and an elementary $j : K \to M$ which is not an iteration map. Assuming only that there is no inner model with a Woodin cardinal, however, we can still show that K is rigid. Using this fact, we can characterize K as the unique universal weasel which is elementarily embeddable in all universal weasels. We shall also show that if $j : K \to M$, where M is $\Omega+1$ iterable, and $\mu = \text{crit}(j)$, then $P(\mu)^K = P(\mu)^M$. We shall assume throughout this section that K^c satisfies "There are no Woodin cardinals", so that Ω is A_0-thick in K^c and $K = \text{Def}(K^c, A_0)$. Since we need only consider S-thick sets for $S = A_0$, we make the following definition.

Definition 8.1. *We say Γ is thick in W iff Γ is A_0-thick in W. Similarly, W has the hull (resp. definability) property at α iff W has the A_0-hull (resp. definability) property at α. Finally, $Def(W) = Def(W, A_0)$.*

We begin by showing that for any $\alpha < \Omega$, one can generate a witness that $\mathcal{J}_\alpha^K$ is A_0-sound from K by taking ultrapowers by the order zero measures at each measurable cardinal κ of K such that $\alpha < \kappa < \Omega$. The key to this result is the following.

Lemma 8.2. *Let W be an $\Omega+1$ iterable weasel which has the hull property at all $\alpha < \Omega$; then there is an iteration tree $\mathcal{T}$ on K with last model $\mathcal{M}_\theta^{\mathcal{T}} = W$, and such that*

(1) $\forall\alpha(\alpha+1<\theta \Rightarrow E_\alpha^{\mathcal{T}}$ *is a normal measure (i.e., has only one generator), so that $\mathcal{T}$ is linear;*

(2) $i_{0,\theta}^{\mathcal{T}}{}''K = Def(W)$.

Proof. Let $\mathcal{T}$ on K and $\mathcal{U}$ on W be the iteration trees resulting from a successful coiteration of K with W determined by the unique $\Omega+1$ iteration strategies on the two weasels. We show first that W never moves; i.e., $\mathcal{U}$ is trivial.

Claim 1. $lh\,\mathcal{U}=1$.

Proof. Assume not, so that $E_0^{\mathcal{U}}$ exists. Let $\alpha<\Omega$ be inaccessible and such that $lh(E_0^{\mathcal{U}})<\alpha$. Let R be an $\Omega+1$ iterable weasel which witnesses that $\mathcal{J}_\alpha^K$ is A_0-sound. Let $\mathcal{S}$ on R and $\mathcal{V}$ on W be the iteration trees resulting from a successful coiteration of R with W. Let $Q=\mathcal{M}_\gamma^{\mathcal{S}}=\mathcal{M}_\delta^{\mathcal{V}}$ be the common last model of the two trees, and $i=i_{0\gamma}^{\mathcal{S}}$ and $j=i_{0\delta}^{\mathcal{V}}$ be the iteration maps. Let $\mu=\text{crit}(j)$.

Since $\mathcal{J}_\alpha^R=\mathcal{J}_\alpha^K$ and $lh(E_0^{\mathcal{U}})<\alpha$, we have $E_0^{\mathcal{V}}=E_0^{\mathcal{U}}$. But then $\mu<\nu(E_0^{\mathcal{V}})<\alpha$. Since $j:W\to Q$ is an iteration map and W has the hull property everywhere, Q has the hull property at all $\xi\le\mu$. Let

$$\theta=\text{least } \eta\in[0,\gamma]_S \text{ such that } \eta=\gamma \text{ or } \text{crit}(i_{\eta\gamma}^{\mathcal{S}})\ge\mu\,.$$

From example 4.3 and the remark following it, we see that whenever $\eta+1\in[0,\theta]_S$, then $E_\eta^{\mathcal{S}}$ is a normal measure, that is, has $\text{crit}(E_\eta^{\mathcal{S}})$ as its only generator. (Otherwise, Q would fail to have the hull property at $(\kappa^+)^Q$, where $\kappa=\text{crit}(E_\eta^{\mathcal{S}})<\mu$. Since $(\kappa^+)^Q\le\mu$, this is impossible.) But now in any normal iteration tree, a normal measure can only be applied to the model from which it is taken. It follows that $\mathcal{S}\restriction\theta+1$ is just a linear iteration of the normal measures $E_\eta^{\mathcal{S}}$ for $\eta+1\le\theta$.

We claim that $\mathcal{M}_\theta^{\mathcal{S}}$ has the definability property at μ. For let Γ be thick in $\mathcal{M}_\theta^{\mathcal{S}}$; we want $a\in(\mu\cup\Gamma)^{<\omega}$ and a term τ such that $\mu=\tau^{\mathcal{M}_\theta^{\mathcal{S}}}[a]$. Let Δ be the thick class of fixed points of $i_{0,\theta}^{\mathcal{S}}$. Suppose first $i_{0,\theta}^{\mathcal{S}}(\mu)=\mu$; then since R witnesses that $\mathcal{J}_\alpha^K$ is A_0-sound, and $\mu<\alpha$, we can find $a\in(\Gamma\cap\Delta)^{<\omega}$ and τ such that $\tau^R[a]=\mu$. But then $\tau^{\mathcal{M}_\theta^{\mathcal{S}}}[a]=\mu$, as desired. Suppose next that $i_{0,\theta}^{\mathcal{S}}(\mu)>\mu$. The usual representation of iterated ultrapowers gives us a function $f:[\mu]^{<\omega}\to\mu$ and an $a\in[\mu]^{<\omega}$ such that $i_{0,\theta}^{\mathcal{S}}(f)(a)=\mu$. Since R witnesses that $\mathcal{J}_\alpha^K$ is A_0-sound, and $f\in\mathcal{J}_\alpha^K$, we have $b\in(\Gamma\cap\Delta)^{<\omega}$ and τ such that $f=\tau^R[b]$. But then $\mu=\tau^{\mathcal{M}_\theta^{\mathcal{S}}}[b](a)$, which gives the desired definition of μ.

Since $\mu=\text{crit}(j)$, Q does not have the definability property at μ. It follows that $\theta<\gamma$ and $\text{crit}(i_{0,\theta}^{\mathcal{S}})=\mu$. But now the ancient Kunen argument yields a contradiction: let $A\subseteq\mu$ and $A\in\mathcal{M}_\theta^{\mathcal{S}}$. Since $\mathcal{M}_\theta^{\mathcal{S}}$ has the hull property at μ, we can write $A\cap\mu=\tau^{\mathcal{M}_\theta^{\mathcal{S}}}[a]$, where $i_{0,\theta}^{\mathcal{S}}(a)=a=j(a)$. It follows that $i_{0,\theta}^{\mathcal{S}}(A)\cap\nu=j(A)\cap\nu$, where $\nu=\inf\,(i_{0,\theta}^{\mathcal{S}}(\mu),j(\mu))$. This implies that the first extender used in $\mathcal{S}$ on $[\theta, \gamma]_S$ and the first extender used in $\mathcal{V}$ on $[0, \delta]_V$ are compatible, which is a contradiction. This proves claim 1. □

The proof of claim 1 also gives:

Claim 2. $\mathcal{T}$ is a linear iteration of normal measures.

Proof. If not, then we can find a weasel R which witnesses that $\mathcal{J}_\alpha^K$ is A_0-sound, where α is inaccessible and large enough that some extender with more than one generator used on $\mathcal{T}$ has length $< \alpha$. Again, let $\mathcal{S}$ on R and $\mathcal{V}$ on W come from coiteration, with $\mathcal{M}_\gamma^{\mathcal{S}} = \mathcal{M}_\delta^{\mathcal{V}}$. Since $\mathcal{J}_\alpha^K = \mathcal{J}_\alpha^R$, $\mathcal{S}$ and $\mathcal{T}$ have the same initial segment using extenders of length $< \alpha$. It follows that there is an $\eta + 1\ S\gamma$ such that $\text{crit}(E_\eta^{\mathcal{S}}) < \alpha$ and $E_\eta^{\mathcal{S}}$ has more than one generator. Thus there is $\xi < \alpha$ such that $\mathcal{M}_\gamma^{\mathcal{S}}$ fails to have the hull property at ξ. On the other hand, the proof of claim 1 shows that $\text{crit}(i_{0,\delta}^{\mathcal{V}}) \geq \alpha$, and since W has the hull property everywhere, this means $\mathcal{M}_\delta^{\mathcal{V}}$ has the hull property at all $\xi \leq \alpha$. This is a contradiction. □

Let $lh\ \mathcal{T} = \theta + 1$, so that $W = \mathcal{M}_\theta^{\mathcal{T}}$.

Claim 3. $\text{Def}(W) = i_{0,\theta}^{\mathcal{T}}{}''K$.

Proof. Let $\alpha < \Omega$ be inaccessible and $i_{0,\theta}^{\mathcal{T}}{}''\alpha \subseteq \alpha$; we shall show that $\text{Def}(W) \cap \alpha = i_{0,\theta}^{\mathcal{T}}{}''\alpha$. Let R witness that $\mathcal{J}_\alpha^K$ is A_0-sound, and let $\mathcal{S}$ on R and $\mathcal{V}$ on W come from coiteration, with $\mathcal{M}_\gamma^{\mathcal{S}} = \mathcal{M}_\delta^{\mathcal{V}}$. Let η be least such that $\eta = \theta$ or $\text{crit}(i_{\eta\theta}^{\mathcal{T}}) \geq \alpha$. Since $\mathcal{J}_\alpha^K = \mathcal{J}_\alpha^R$, $\mathcal{S} \restriction \eta + 1 = \mathcal{T} \restriction \eta + 1$. We claim that $\eta + 1 \subseteq [0, \gamma]_S$; that is, the linear iteration giving us $\mathcal{T} \restriction \eta + 1$ is an initial segment of the branch of $\mathcal{S}$ leading to $\mathcal{M}_\gamma^{\mathcal{S}}$. For otherwise, letting $\beta + 1 \in [0, \gamma]_S$ be least such that $lh\ E_\beta^{\mathcal{S}} \geq \alpha$, we have $\text{crit}\ E_\beta^{\mathcal{S}} < \alpha$. This implies that $\mathcal{M}_\gamma^{\mathcal{S}}$ has the hull property at all $\xi \leq \text{crit}(E_\beta^{\mathcal{S}})$, but not at $(\text{crit}(E_\beta^{\mathcal{S}})^+)^{\mathcal{M}_\gamma^{\mathcal{S}}}$. But then we get $\text{crit}(i_{0,\delta}^{\mathcal{V}}) = \text{crit}(E_\beta^{\mathcal{S}})$, and that $E_\beta^{\mathcal{S}}$ is compatible with the first extender used in $[0, \delta]_V$, as in the ancient Kunen argument. This is a contradiction.

Thus $\eta + 1 \subseteq [0, \gamma]_S$, and the argument also shows $\text{crit}(i_{\eta\gamma}^{\mathcal{S}}) \geq \alpha$. So

$$i_{0,\theta}^{\mathcal{T}} \restriction \alpha = i_{0,\eta}^{\mathcal{T}} \restriction \alpha = i_{0,\eta}^{\mathcal{S}} \restriction \alpha = i_{0,\gamma}^{\mathcal{S}} \restriction \alpha .$$

Finally, since $\mathcal{M}_\delta^{\mathcal{V}}$ has the hull property everywhere below α, $\text{crit}(i_{0\delta}^{\mathcal{V}}) \geq \alpha$. This gives

$$\begin{aligned} \text{Def}(W) \cap \alpha &= \text{Def}(\mathcal{M}_\delta^{\mathcal{V}}) \cap \alpha = Def(\mathcal{M}_\gamma^{\mathcal{S}}) \cap \alpha \\ &= i_{0\gamma}^{\mathcal{S}}{}''(\text{Def}(R) \cap \alpha) = i_{0\gamma}^{\mathcal{S}}{}''\alpha \\ &= i_{0,\theta}^{\mathcal{T}}{}''\alpha . \qquad \square \end{aligned}$$

Clearly, the claims yield 8.2. □

Lemma 8.3. *Suppose $K^c \models$ "There are no Woodin cardinals". Let α be a cardinal of K, and let W be the iterate of K obtained by taking ultrapowers by* *the order zero total measure at each measurable cardinal of K which is $\geq \alpha$. Then W witnesses that $\mathcal{J}^K_\alpha$ is A_0-sound; moreover, W has the hull property at all $\beta < \Omega$.*

Proof. Clearly W is universal, and no $\kappa \in A_0$ is measurable in W. It follows from 3.7(1) that Ω is thick in W, and it remains only to show that $\alpha \subseteq \text{Def}(W)$ and that W has the hull property everywhere. Let $i : K \to W$ be the iteration map.

Lemma 4.5 gives us an $\Omega+1$-iterable weasel M which has the hull property at all $\beta < \Omega$. Let α^* be the αth ordinal in Def(M), and for each $\beta < \alpha^*$ such that $\beta \notin \text{Def}(M)$, pick a thick class Γ_β such that $\beta \notin H^M(\Gamma_\beta)$.

Let

$$R = \text{transitive collapse of } H^M(\bigcap\{\Gamma_\beta \mid \beta \in \alpha^* - \text{Def}(M)\}).$$

Clearly, $\alpha \subseteq \text{Def}(R)$, so R witnesses that $\mathcal{J}^K_\alpha$ is A_0-sound. It is easy to see that R has the hull property at all $\beta < \Omega$. [Let $A \subseteq \beta$, $A \in R$, and let Γ be thick in R. Let $\pi : R \to M$ be the collapse map. Since $\pi''\Gamma$ is thick in M, and M has the hull property at $\pi(\beta)$, there is a term τ and $b \in \Gamma^{<\omega}$ and $a \in \pi(\beta)^{<\omega}$ such that $\pi(A) = \tau^M[a, \pi(b)] \cap \pi(\beta)$. The least a with this property is of the form $\pi(\bar{a})$ for some $\bar{a} \in \beta^{<\omega}$, and then $A = \tau^R[\bar{a}, b] \cap \beta$.]

Let $j : K \to R$ be the iteration map associated to the linear iteration of normal measures given by 8.2. We have $j''K = \text{Def}(R)$, and hence $\alpha \leq \text{crit}(j)$.

Let $\mathcal{T}$ on R and $\mathcal{U}$ on W be the iteration trees associated to a successful coiteration of R with W, and let $\mathcal{M}^{\mathcal{T}}_\gamma = \mathcal{M}^{\mathcal{U}}_\delta$ be the common last model. Since $i^{\mathcal{T}}_{0\gamma} \circ j$ and $i^{\mathcal{U}}_{0\delta} \circ i$ are iteration maps, the Dodd-Jensen lemma gives $i^{\mathcal{T}}_{0\gamma} \circ j = i^{\mathcal{U}}_{0\delta} \circ i$. From this we get that $\text{crit}(i^{\mathcal{T}}_{0\gamma}) \geq \alpha$ and $\text{crit}(i^{\mathcal{U}}_{0\delta}) \geq \alpha$. For otherwise, since i and j have critical point $\geq \alpha$, we get $\kappa < \alpha$ such that $\kappa = \text{crit}(i^{\mathcal{T}}_{0\gamma}) = \text{crit}(i^{\mathcal{U}}_{0\delta})$. Also, for any $A \subseteq \kappa$ such that $A \in K$, $i(A) = j(A) = A$, so $i^{\mathcal{T}}_{0\gamma}(A) = i^{\mathcal{U}}_{0\delta}(A)$. This leads to the usual contradiction that the first extenders used in $[0, \gamma]_T$ and $[0, \delta]_U$ are compatible.

Now by 5.5, $i^{\mathcal{T}}_{0\gamma}{''}\text{Def}(R) = \text{Def}(\mathcal{M}^{\mathcal{T}}_\gamma) = \text{Def}(\mathcal{M}^{\mathcal{U}}_\delta) = i^{\mathcal{U}}_{0\delta}{''}\text{Def}(W)$. Since $\alpha \subseteq \text{Def}(R)$ and $\text{crit}(i^{\mathcal{T}}_{0\gamma}) \geq \alpha$, $\alpha \subseteq \text{Def}(W)$.

It remains to show that W has the hull property everywhere. Let $\beta < \Omega$ be a cardinal of K, with $\alpha < \beta$. Let W^* be the iterate of K obtained by hitting the order zero measure on each measurable cardinal $\kappa \geq \beta$ of K exactly once, so that W^* witnesses that $\mathcal{J}^K_\beta$ is A_0-sound by what we have just shown. In particular, W^* has the hull property at all $\gamma < \beta$. Clearly, W is a linear iterate of W^* by normal measures (i.e., those of order zero on cardinals κ such that $\alpha \leq \kappa < \beta$), and therefore W has the hull property at all $\gamma < \beta$. Since β was arbitrary, W has the hull property everywhere. □

The reader may have noticed that 4.5 and 8.2 gave us a linear iterate by normal measures W of K satisfying the conclusion of 8.3, without much effort. (See paragraph 2 of the proof of 8.3.) What 8.3 gives, beyond this, is an iteration leading from K to W which is definable over K.

Our next lemma expresses a maximality property of K and its iterates.

Definition 8.4. *Let $\mathcal{M}$ be a premouse. We say that an extender F coheres with $\mathcal{M}$ just in case $(J_\alpha^{\mathcal{M}}, \in, \vec{E}^{\mathcal{M}} \restriction \alpha, \tilde{F})$ is a premouse, where $\alpha = lh\ F$.*

Of course, any extender on the $\mathcal{M}$-sequence coheres with $\mathcal{M}$. As another example: if $(\mathcal{T}, \mathcal{U})$ is a coiteration in which, at some intermediate stage, the current models are $\mathcal{M}_\alpha^{\mathcal{T}}$ and $\mathcal{M}_\beta^{\mathcal{U}}$, and $E_\beta^{\mathcal{U}}$ is part of the least disagreement at this stage, then $E_\beta^{\mathcal{U}}$ coheres with $\mathcal{M}_\alpha^{\mathcal{T}}$.

We now show that if an extender E coheres with the last model $\mathcal{M}$ of an iteration tree $\mathcal{T}$ on K, and a certain iterability condition is satisfied, then E is on the $\mathcal{M}$-sequence. The iterability condition is that we can extend $\mathcal{T}$ one step using E as if it came from the $\mathcal{M}$-sequence, and then continue iterating in the normal fashion. The following definition enables us to make this condition precise. The reader should see 9.6 and 9.7 for the general notion of a phalanx, and the definition of $\Phi(\mathcal{T})$, the phalanx derived from an iteration tree $\mathcal{T}$.

Definition 8.5. *Let T be an iteration tree with last model $\mathcal{M}_\alpha^{\mathcal{T}}$, let E cohere with $\mathcal{M}_\alpha^{\mathcal{T}}$, and suppose $lh\ E > lh\ E_\beta^{\mathcal{T}}$ for all $\beta < \alpha$. Let γ be least such that $\gamma = \alpha$ or $crit(E) < \nu(E_\gamma^{\mathcal{T}})$, and let $\mathcal{P}$ be the longest initial segment $\mathcal{R}$ of $\mathcal{M}_\gamma^{\mathcal{T}}$ such that $P(\kappa) \cap \mathcal{R} = P(\kappa) \cap \mathcal{J}_{lhE}^{\mathcal{M}_\alpha^{\mathcal{T}}}$, where $\kappa = crit(E)$. Let $k < \omega$ be least such that $\rho_{k+1}(\mathcal{P}) \leq \kappa$, if there is such a $k < \omega$, and let $k = \omega$ otherwise. Suppose $\mathcal{N} = Ult_k(\mathcal{P}, E)$ is wellfounded. Then the E-extension of $\Phi(\mathcal{T})$ is the phalanx $\Phi(\mathcal{T})^\frown\langle \mathcal{N}, k, \nu, \lambda\rangle$, where $\nu = \nu(E)$ and λ is the least cardinal of $\mathcal{N}$ which is $\geq \nu$.*

Theorem 8.6. *Suppose $K^c \models$ There are no Woodin cardinals". Let T be a normal iteration tree on W of length $\alpha + 1 < \Omega$, and suppose E coheres with $\mathcal{M}_\alpha^{\mathcal{T}}$ and $lh\ E \geq lhE_\beta^{\mathcal{T}}$ for all $\beta < \alpha$. Suppose that W witnesses that $\mathcal{J}_\mu^K$ is A_0-sound, where μ is inaccessible and large enough that $\alpha < \mu$ and $E \in V_\mu$ and $\forall\beta < \alpha(E_\beta^{\mathcal{T}} \in V_\mu)$. The following are equivalent:*

(a) *E is on the $\mathcal{M}_\alpha^{\mathcal{T}}$ sequence,*

(b) *the E-extension of $\Phi(\mathcal{T})$ is $\Omega + 1$ iterable.*

Proof. (a) $\Rightarrow$ (b) is just a re-statement of the fact that W is $\Omega + 1$ iterable. Now suppose $\mathcal{B} = \Phi(\mathcal{T}^\frown\langle \mathcal{N}, k, \nu, \lambda\rangle$ is the E-extension of $\Phi(\mathcal{T})$, and that $\mathcal{B}$ is Ω+1 iterable. Let us form the natural coiteration of $\mathcal{B}$ with $\Phi(\mathcal{T})$: at successor steps we iterate the least disagreement, beginning with the least disagreement between the last models $\mathcal{N}$ of $\mathcal{B}$ and $\mathcal{M}_\alpha^{\mathcal{T}}$ of $\Phi(\mathcal{T})$; the rules for iteration trees on phalanxes determine the models to which we apply the extenders from the least disagreement. At limit steps we use the (unique) $\Omega + 1$ iteration strategies for $\mathcal{B}$ and $\Phi(\mathcal{T})$ to pick branches. The usual argument shows that this coiteration terminates successfully at some stage $\leq \Omega$. The iteration tree on $\Phi(\mathcal{T})$ which is produced can clearly be regarded as an extension of $\mathcal{T}$; let us call this extension $\mathcal{U}$. We note that $\mathcal{U}$ is normal. [This is clear, except perhaps for the increasing-length condition on its extenders; for that we need only show $lh\ E^{\mathcal{U}}_{\alpha} > lh\ E^{\mathcal{U}}_{\beta}$ for all $\beta < \alpha$. But since E coheres with $\mathcal{M}^{\mathcal{U}}_{\alpha} = \mathcal{M}^{\mathcal{T}}_{\alpha}$, $\mathcal{N}$ agrees with $\mathcal{M}^{\mathcal{U}}_{\alpha}$ below $lh(E)$, and the disagreement of which $E^{\mathcal{U}}_{\alpha}$ is a part must occur at a length $\geq lh(E)$.] The iteration tree on $\mathcal{B}$ which is produced can be regarded as a system $\mathcal{S}$ extending $\mathcal{T}$ which has all the properties of a normal iteration tree except that, since $E^{\mathcal{S}}_{\alpha} = E$, it may not be true that $E^{\mathcal{S}}_{\alpha}$ is on the $\mathcal{M}^{\mathcal{S}}_{\alpha}$ sequence. We shall use iteration tree terminology in connection with $\mathcal{S}$; its meaning should be clear.

Claim 1. If $E^{\mathcal{S}}_{\eta}$ is compatible with $E^{\mathcal{U}}_{\xi}$, then $\eta = \xi \leq \alpha$ and $E^{\mathcal{S}}_{\eta} = E^{\mathcal{U}}_{\xi}$.

Proof. If $\mathcal{V}$ is a normal iteration tree, then $lh\ E^{\mathcal{V}}_{\sigma}$ is a cardinal in all models $\mathcal{M}^{\mathcal{V}}_{\tau}$ for $\tau > \sigma$, so that $E^{\mathcal{V}}_{\sigma}$ is incompatible with any extender on the $\mathcal{M}^{\mathcal{V}}_{\tau}$-sequence, for $\tau > \sigma$, by the initial segment condition and the fact that $E^{\mathcal{V}}_{\sigma}$ collapses its length. The system $\mathcal{S}$ has this property of normal trees because $E^{\mathcal{S}}_{\alpha}$ coheres with $\mathcal{M}^{\mathcal{S}}_{\alpha}$. Since $\mathcal{S} \upharpoonright \alpha = \mathcal{U} \upharpoonright \alpha = \mathcal{T}$, this gives the claim immediately if $\eta < \alpha$ or $\xi < \alpha$.

It cannot happen that $\eta > \alpha$ and $\xi \geq \alpha$. For then $E^{\mathcal{S}}_{\eta}$ and $E^{\mathcal{U}}_{\xi}$ are each part of a disagreement in the coiteration of $\mathcal{B}$ with $\Phi(\mathcal{T})$. They cannot be part of the same disagreement since they are compatible. Thus $lh(E^{\mathcal{S}}_{\eta}) \neq lh(E^{\mathcal{U}}_{\xi})$. Suppose that $lh(E^{\mathcal{S}}_{\eta}) < lh(E^{\mathcal{U}}_{\xi})$; the other case leads to a similar contradiction. By the inital segment condition on premice, $E^{\mathcal{S}}_{\eta}$ is on the $\mathcal{M}^{\mathcal{U}}_{\xi}$ sequence or an ultrapower thereof. Letting $\mathcal{M}^{\mathcal{S}}_{\tau}$ be the model on $\mathcal{S}$ with which $\mathcal{M}^{\mathcal{U}}_{\xi}$ is being compared, we have $\eta < \tau$ and $E^{\mathcal{S}}_{\eta}$ on the $\mathcal{M}^{\mathcal{S}}_{\tau}$ sequence or an ultrapower thereof. This means $lh(E^{\mathcal{S}}_{\eta})$ is not a cardinal of $\mathcal{M}^{\mathcal{S}}_{\tau}$, a contradiction.

We are left with the possibility that $\eta = \alpha$ and $\xi \geq \alpha$. Thus $E^{\mathcal{S}}_{\eta} = E$. If $lh(E^{\mathcal{U}}_{\xi}) > lh(E)$, we get that E is on the $\mathcal{M}^{\mathcal{U}}_{\xi}$ sequence or an ultrapower thereof, and hence on the $\mathcal{N}$-sequence or an ultrapower thereof, which is impossible since E coheres with $\mathcal{M}^{\mathcal{T}}_{\alpha}$. Thus $E^{\mathcal{U}}_{\xi} = E^{\mathcal{S}}_{\eta}$. □

Now let $\mathcal{M}^{\mathcal{S}}_{\gamma}$ and $\mathcal{M}^{\mathcal{U}}_{\delta}$ be the last models of $\mathcal{S}$ and $\mathcal{U}$ respectively.

Claim 2. $\mathcal{M}^{\mathcal{S}}_{\gamma} = \mathcal{M}^{\mathcal{U}}_{\delta}$.

Proof. If $[0,\gamma]_S \cap D^{\mathcal{S}} = \phi$, then since Ω is thick in $W = \mathcal{M}^{\mathcal{S}}_0$, Ω is thick in $\mathcal{M}^{\mathcal{S}}_{\gamma}$. But then $[0,\delta]_U \cap D^{\mathcal{U}} = \phi$ and hence $\mathcal{M}^{\mathcal{U}}_{\delta} = \mathcal{M}^{\mathcal{S}}_{\gamma}$. If $[0,\gamma]_S \cap D^{\mathcal{S}} \neq \phi$, then $\mathcal{M}^{\mathcal{U}}_{\delta} \trianglelefteq \mathcal{M}^{\mathcal{S}}_{\gamma}$ because $\mathcal{M}^{\mathcal{S}}_{\gamma}$ is not ω-sound. But then $[0,\delta]_U \cap D^{\mathcal{U}} \neq \phi$, as otherwise Ω is thick in $\mathcal{M}^{\mathcal{U}}_{\delta}$ while $\mathcal{M}^{\mathcal{S}}_{\gamma}$ computes κ^+ incorrectly for a.e. $\kappa \in A_0$. Thus $\mathcal{M}^{\mathcal{S}}_{\gamma} \trianglelefteq \mathcal{M}^{\mathcal{U}}_{\delta}$, as $\mathcal{M}^{\mathcal{U}}_{\delta}$ is not ω-sound. □

Now let $\theta \leq \alpha$ be largest such that $\theta S \gamma$ and $\theta U \delta$. Since $\mathcal{M}^{\mathcal{S}}_{\alpha+1} = \mathcal{N}$ exists, $\theta < \gamma$, so we can set $\eta + 1 =$ unique $\beta \in [0,\gamma]_S$ such that S-pred$(\beta) = \theta$.

Claim 3. $\theta < \delta$.

Proof. If not, then since $\theta \leq \alpha \leq \delta$, we must have $\theta = \alpha = \delta$.

Suppose first that $D^{\mathcal{T}} \cap [0,\alpha]_T = \phi$. Since then $\mathcal{M}_\alpha^{\mathcal{T}} = \mathcal{M}_\gamma^{\mathcal{S}}$ is a universal weasel, $[0,\gamma]_S \cap D^{\mathcal{S}} = \phi$. Let $\kappa = \text{crit}(E_\eta^{\mathcal{S}})$. Since $\kappa < \nu(E_\theta) < \mu$, and $\mu \subseteq \text{Def}(W)$, $\mathcal{M}_\theta^{\mathcal{S}}$ has the definability property at κ. Since $\kappa = \text{crit } i_{\theta,\gamma}^{\mathcal{S}}$, $\mathcal{M}_\gamma^{\mathcal{S}}$ does not have the definability property at κ. But $\mathcal{M}_\theta^{\mathcal{S}} = \mathcal{M}_\alpha^{\mathcal{T}} = \mathcal{M}_\delta^{\mathcal{U}} = \mathcal{M}_\delta^{\mathcal{S}}$, a contradiction.

By claim 3, we can let $\xi + 1 \in [0,\delta]_U$ be the unique β such that U-pred$(\beta) = \theta$.

Claim 4. $E_\eta^{\mathcal{S}}$ is compatible with $E_\xi^{\mathcal{U}}$.

Proof. We claim first that $D^{\mathcal{S}} \cap (\eta+1,\gamma]_S = \phi$, and $\deg^{\mathcal{S}}(\eta+1) = \deg^{\mathcal{S}}(\gamma)$. For otherwise, let $\sigma+1$ be the site of the last drop in model or degree along $(\eta+1,\gamma]_S$, and let $k = \deg^{\mathcal{S}}(\sigma+1)$. By standard arguments, $(\mathcal{M}_{\sigma+1}^*)^{\mathcal{S}} = \mathfrak{C}_{k+1}(\mathcal{M}_\gamma^{\mathcal{S}})$, k is least such that $\mathcal{M}_\gamma^{\mathcal{S}}$ is not $k+1$ sound, and $i_{\sigma+1,\gamma}^{\mathcal{S}} \circ (i_{\sigma+1}^*)^{\mathcal{S}}$ is the canonical core embedding from $\mathfrak{C}_{k+1}(\mathcal{M}_\gamma^{\mathcal{S}})$ into $\mathcal{M}_\gamma^{\mathcal{S}}$. Since $\mathcal{M}_\gamma^{\mathcal{S}} = \mathcal{M}_\delta^{\mathcal{U}}$, there is a last drop $\tau+1$ in model or degree along $[0,\delta]_U$, and $\deg^{\mathcal{U}}(\tau+1) = k$, $(\mathcal{M}_{\tau+1}^*)^{\mathcal{U}} = \mathfrak{C}_{k+1}(\mathcal{M}_\delta^{\mathcal{U}})$, and $i_{\tau+1,\delta}^{\mathcal{U}} \circ (i_{\tau+1}^*)^{\mathcal{U}}$ is the core embedding. This gives $E_\sigma^{\mathcal{S}}$ compatible with $E_\tau^{\mathcal{U}}$, so that $\sigma = \tau \leq \alpha$ by claim 1. But then $\sigma \leq \theta$ by the definition of θ, while $\theta \leq \eta < \sigma$ by the definition of σ.

A similar argument shows that $D^{\mathcal{U}} \cap (\xi+1,\delta)_U = \phi$ and $\deg^U(\xi+1) = \deg^U(\gamma)$.

Next, suppose $\eta+1 \in D^{\mathcal{S}}$ or $\deg^{\mathcal{S}}(\eta+1) \neq \deg^{\mathcal{S}}(\theta)$. Arguing as above, we get that $E_\eta^{\mathcal{S}}$ is compatible with $E_\tau^{\mathcal{U}}$, where $\tau+1$ is the site of the last drop in model or degree along $[0,\delta]_U$. We cannot have $\tau+1 \in [0,\theta]_U$, since then $E_\tau^{\mathcal{U}} = E_\tau^{\mathcal{S}}$, so $E_\tau^{\mathcal{S}}$ is compatible with $E_\eta^{\mathcal{S}}$, while $\tau \neq \eta$ because $\eta+1 \notin [0,\theta]_U$. The only other possibility is $\tau+1 = \xi+1$, which gives $E_\eta^{\mathcal{S}}$ compatible with $E_\xi^{\mathcal{U}}$, as desired.

Similarly, if $\xi+1 \in D^{\mathcal{U}}$ or $\deg^{\mathcal{U}}(\xi+1) \notin \deg^{\mathcal{U}}(\theta)$, then $E_\eta^{\mathcal{S}}$ is compatible with $E_\xi^{\mathcal{U}}$. So we may assume that $i_{\theta,\gamma}^{\mathcal{S}}$ and $i_{\theta,\delta}^{\mathcal{U}}$ exist, and each is a $\deg^{\mathcal{T}}(\theta)$ embedding.

Suppose that $D^{\mathcal{T}} \cap [0,\theta]_T = \phi$, and let $\nu = \sup\{\nu(E_\sigma^{\mathcal{T}}) \mid \sigma+1 \in [0,\theta]_T\}$. Since $\mu \subseteq \text{Def}(W)$, we have $\mu \subseteq H^{\mathcal{M}_\theta^{\mathcal{T}}}(\nu \cup \Gamma)$ for all thick Γ. Taking Γ to be the class of common fixed points of $i_{\theta,\gamma}^{\mathcal{S}}$ and $i_{\theta,\delta}^{\mathcal{U}}$, and noting that $\nu < \text{crit}(i_{\theta,\gamma}^{\mathcal{S}}) < \mu$ and $\nu < \text{crit}(i_{\theta,\delta}^{\mathcal{U}}) < \mu$, we get that $i_{\theta,\gamma}^{\mathcal{S}}(A) = i_{\theta,\delta}^{\mathcal{U}}(A)$ for all $A \subseteq \kappa$, where κ is the common critical point of the two embeddings. This implies othat $E_\eta^{\mathcal{S}}$ is compatible with $E_\xi^{\mathcal{U}}$.

Finally, suppose that $D^{\mathcal{T}} \cap [0,\theta]_T \neq \phi$, and again let $\nu = \sup\{\nu(E_\sigma^{\mathcal{T}}) \mid \sigma+1 \in [0,\theta]_T\}$. Let $k = \deg^{\mathcal{T}}(\theta)$. Then $\mathcal{M}_\theta^{\mathcal{T}} = H_{k+1}^{\mathcal{M}_\theta^{\mathcal{T}}}(\nu \cup p)$, where $p = p_{k+1}(\mathcal{M}_\theta^{\mathcal{T}})$. Since $i_{\theta,\gamma}^{\mathcal{S}}(p) = p_{k+1}(\mathcal{M}_\gamma^{\mathcal{S}}) = i_{\theta,\delta}^{\mathcal{U}}(p)$, we again get $i_{\theta,\gamma}^{\mathcal{S}}(A) = i_{\theta,\delta}^{\mathcal{U}}(A)$ for all A contained in the common critical point of the two embeddings, and thus that $E_\eta^{\mathcal{S}}$ is compatible with $E_\xi^{\mathcal{U}}$. This proves claim 4. □

By claims 1 and 4, $\eta = \xi \leq \alpha$ and $E^{\mathcal{S}}_{\eta} = E^{\mathcal{U}}_{\xi}$. We cannot have $\eta < \alpha$, for then $\eta + 1 = \xi + 1 \leq \alpha$, which gives $\eta + 1 \leq \theta$, contrary to the definition of η. Thus $\eta = \xi = \alpha$, so that $E = E^{\mathcal{S}}_{\eta} = E^{\mathcal{U}}_{\alpha}$. Since $E^{\mathcal{U}}_{\alpha}$ is on the $\mathcal{M}^{\mathcal{T}}_{\alpha}$ sequence, we are done. □

Remark 8.7. (a) We believe that one can prove 8.6 with K replacing W. That is, suppose $K^c \models$ "There are no Woodin cardinals", and let $\mathcal{T}$ be a normal iteration tree of length $\alpha + 1 < \Omega$. suppose E coheres with $\mathcal{M}^{\mathcal{T}}_{\alpha}$, $lh(E) > lh(E^{\mathcal{T}}_{\beta})$ for all $\beta < \alpha$, and the E-extension of $\Phi(\mathcal{T})$ is $\Omega+1$-iterable. Then E is on the $\mathcal{M}^{\mathcal{T}}_{\alpha}$ sequence. For let μ be inaccessible and $\mathcal{T}$, $E \in V_{\mu}$, and let W come from K by iterating normal order zero measures above μ, as in 8.3. Let $\pi : K \to W$ be the iteration map, and let $\pi\mathcal{T}$ on W be the copied tree; it is enough to show E is on the $\mathcal{M}^{\pi\mathcal{T}}_{\alpha}$ sequence. By 8.6, it is enough for this to show that the E-extension of $\Phi(\pi\mathcal{T})$ is $\Omega+1$-iterable. We believe that one can do this by chasing the proper diagrams, but haven't gone through the details.

(b) We believe, but haven't checked carefully, that the methods of §9 show that if $\mathcal{T}$, E, and α are as described in the hypothesis of 8.6, and if $(\mathcal{M}^{\mathcal{T}}_{\alpha}, E)$ is countably certified (in the sense of 1.2), then the E-extension of $\Phi(\mathcal{T})$ is iterable. Taking $\mathcal{T} = \emptyset$, this means that every countably certified extender which coheres with K is on the K-sequence.

We now show that K is rigid.

Theorem 8.8. *Suppose that $K^c \models$ "There are no Woodin cardinals", and let $j : K \to K$ be elementary; then $j =$ identity.*

Proof. Suppose otherwise, and let $\kappa = \mathrm{crit}(j)$. Let $\mu < \Omega$ be inaccessible and such that $j(\kappa) < \mu$. Let W be the result of hitting each order zero measure of K with critical point $> \mu$ exactly once (in increasing order) as in 8.3. Thus W witnesses that $\mathcal{J}^K_{\mu}$ is A_0-sound. Let F be the length $j(\kappa)$ extender over K, or equivalently W, derived from j. It will be enough to show $F \restriction \rho \in W$ for all $\rho < j(\kappa)$, since then these initial segments of F witness that κ is Shelah in W. The proof of this is an induction on ρ organized as is the proof of Lemma 11.4 of [FSIT].

Lemma 8.9. *Let $(\kappa^+)^W \leq \rho < j(\kappa)$, and suppose ρ is the sup of the generators of $F \restriction \rho$. Let G be the trivial completion of $F \restriction \rho$, and $\gamma = lh\ G$. Then $\dot{E}^W_{\gamma} = G = F \restriction \gamma$ unless ρ is a limit ordinal greater than $(\kappa^+)^W$, and is itself a generator of F. In this case*

$$G = \begin{cases} E_{\gamma} & \text{if } \rho \notin \mathrm{dom}\ \dot{E}^W, \\ i(\dot{E}^W)_{\gamma} & \text{if } \rho \in \mathrm{dom}\ \dot{E}^W, \end{cases}$$

where $i : \mathcal{J}^W_{\rho} \to Ult_0(\mathcal{J}^W_{\rho}, \dot{E}^W_{\rho})$ is the canonical embedding.

Proof. By induction on ρ. Suppose first that ρ is not a generator of F which is $> (\kappa^+)^W$. It follows that the natural embedding from $\mathrm{Ult}(W, F \restriction \rho)$ into

Ult(W, F) has critical point at least $\gamma = (\rho^+)^{\mathrm{Ult}(W, F\restriction\rho)}$. From this we get that G coheres with W; $(J_\gamma^W, \in, \dot{E}^W, \tilde{G})$ satisfies the initial segment condition on premice by our induction hypothesis. We now apply 8.6 to the trivial iteration tree on W whose only model is W. By 8.6, we are done if we show that the phalanx $\mathcal{B} = (\langle (W, \omega), (\mathrm{Ult}(W, G), \omega)\rangle, \langle \rho, \gamma\rangle)$ is $\Omega + 1$ -iterable, for then G is on the W sequence.

In order to show that $\mathcal{B}$ is $\Omega+1$ iterable, it is enough to find an elementary $\pi : \mathrm{Ult}(W, G) \to W$ such that $\pi \restriction \rho =$ identity, since then we can copy iteration trees on $\mathcal{B}$ as ordinary iteration trees on W. Now W is definable over K from μ, hence $j : W \to j(W)$ is elementary, where $j(W)$ is defined over K from $j(\mu)$ as W was from μ. Since G is an initial segment of the extender derived from j, we have the diagram

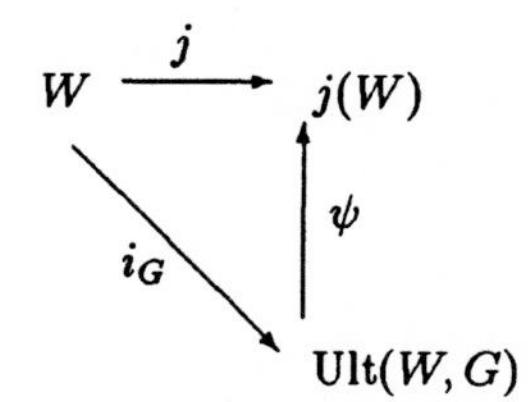

where $\psi \restriction \rho =$ identity. So it is enough to find an embedding $\sigma : j(W) \to W$ which is the identity up to ρ.

Let $\mathcal{U}$ be the linear iteration leading from K to W, so that $K = \mathcal{M}_0^{\mathcal{U}}$ and $W = \mathcal{M}_\Omega^{\mathcal{U}}$. Let $\mathcal{S} = j(\mathcal{U})$ be the linear iteration leading from K to $j(W)$, so that $K = \mathcal{M}_0^{\mathcal{S}}$ and $j(W) = \mathcal{M}_\Omega^{\mathcal{S}}$. Let $\alpha < \Omega$ be least so that $\mathrm{crit}(i_{\alpha,\Omega}^{\mathcal{U}}) > j(\mu)$. We now define by induction on $\beta \geq 0$ maps $\tau_\beta : \mathcal{M}_\beta^{\mathcal{S}} \to \mathcal{M}_{\alpha+\beta}^{\mathcal{U}}$ such that $\tau_\gamma \circ i_{\beta\gamma}^{\mathcal{S}} = i_{\alpha+\beta,\alpha+\gamma}^{\mathcal{U}}$ for $\beta < \gamma$. We begin with $\tau_0 = i_{0,\alpha}^{\mathcal{U}}$. Given τ_β, set

$$\tau_{\beta+1}([f]_H) = [\tau_\beta(f)]_{\tau_\beta(H)},$$

where $H = E_\beta^{\mathcal{S}}$; notice here that $\tau_\beta(H) = E_{\alpha+\beta}^{\mathcal{U}}$ so that this works out. [Let Z be the $\alpha + \beta$th order zero total measure of K above μ. By definition of α, Z is the βth order zero total measure of K above $j(\mu)$. Thus $E_{\alpha+\beta}^{\mathcal{U}} = i_{0,\alpha+\beta}^{\mathcal{U}}(Z)$ and $E_\beta^{\mathcal{S}} = i_{0\beta}^{\mathcal{S}}(Z)$. But then $\tau_\beta(E_\beta^{\mathcal{S}}) = \tau_\beta(i_{0\beta}^{\mathcal{S}}(Z)) = i_{\alpha,\alpha+\beta}^{\mathcal{U}}(\tau_0(Z)) = i_{0,\alpha+\beta}^{\mathcal{U}}(Z) = E_{\alpha+\beta}^{\mathcal{U}}$.] We define τ_λ using commutativity, for λ a limit. Clearly $\tau_\Omega : j(W) \to W$, and it is easy to see that $\tau_\Omega \restriction \mu$ is the identity.

This finishes the proof of 8.9 in the case $\rho = (\kappa^+)^W$ or ρ is not a generator of F. If $\rho > (\kappa^+)^W$ and ρ is a generator of F, then the natural embedding $\psi : \mathrm{Ult}(W, G) \to j(W)$ has critical point ρ. It is therefore not obvious that G coheres with W. Nevertheless, we can apply the condensation Theorem 8.2 of [FSIT]. For notice that $\psi(\rho) < j(\kappa) < \mu$, and that W and $j(W)$ agree below μ, and that ρ is a cardinal of Ult(W, G). Further, $\gamma = (\rho^+)^{\mathrm{Ult}(W,G)}$, so there are arbitrarily large $\eta < \gamma$ such that $\rho_\omega(\mathcal{J}_\eta^{\mathrm{Ult}(W,G)}) = \rho$ and ψ maps

$\mathcal{J}_\eta^{\mathrm{Ult}(W,G)}$ elementarily into $\mathcal{J}_{\psi(\eta)}^{j(W)}$ which is a level of W. If $\rho \notin \mathrm{dom}\ \dot{E}^W$, then 8.2 of [FSIT] gives $\mathcal{J}_\eta^{\mathrm{Ult}(W,G)} = \mathcal{J}_\eta^W$ for all $\eta < \gamma$, so that G coheres with W. (Again, the initial segment condition is our induction hypothesis.) We can then finish the proof just as in the case ρ is not a generator of F. So assume $\rho \in \mathrm{dom}\ \dot{E}^W$.

In this case, 8.2 of [FSIT] implies that G coheres with $\mathrm{Ult}(W, \dot{E}_\rho^W)$. (Notice that $\mathrm{Ult}(W, \dot{E}_\rho^W)$ makes sense: letting $\lambda = \mathrm{crit}\ \dot{E}_\rho^W$, we have $(\lambda^+)\mathcal{J}_\rho^W = (\lambda^+)^{\mathrm{Ult}(W,G)}$ because ρ is a cardinal of $\mathrm{Ult}(W, G)$, which agrees with W below ρ. Since $\rho = \mathrm{crit}(\psi)$, $(\lambda^+)^{\mathrm{Ult}(W,G)} = (\lambda^+)^{j(W)} = (\lambda^+)^W$. Thus $\dot{E}_\rho^W$ measures all sets in W.) The proof of 8.9 will be complete if we show that G is on the $\mathrm{Ult}(W, \dot{E}_\rho^W)$ sequence. To this end, we apply 8.6 to the iteration tree $\mathcal{T}$ on W of length 2 such that $E_0^{\mathcal{T}} = \dot{E}_\rho^W$. We are done if we can show that the G-extension of $\Phi(\mathcal{T})$ is $\Omega+1$ iterable. Using the natural map from $\mathrm{Ult}(W, G)$ into $j(W)$ we can copy iteration trees on the G-extension of $\Phi(\mathcal{T})$ as iteration trees on $\mathcal{B}$, where $\mathcal{B}$ is the phalanx with models $\langle W, \mathrm{Ult}(W, \dot{E}_\rho^W), j(W)\rangle$ and "exchange ordinals" $\langle \nu(\dot{E}_\rho^W), \rho\rangle$. So it is enough to see that $\mathcal{B}$ is $\Omega+1$-iterable. But we can embed $\mathcal{B}$ into a K^c-generated phalanx as follows: let $\pi : K \to K^c$ with $\mathrm{ran}(\pi) = \mathrm{Def}(K^c)$. Since W is obtained from K by a K-definable iteration process with critical points above μ, we have $\pi : W \to R$, where R is obtained in the same way from K^c, the critical points being above $\pi(\mu)$. Similarly, $\pi : j(W) \to S$ where S is obtained from K^c by iterating above $\pi(j(\mu))$. Finally, let $\sigma : \mathrm{Ult}(W, \dot{E}_\rho^W) \to \mathrm{Ult}(R, \pi(\dot{E}_\rho^W))$ be the shift map induced by π. We have $\sigma \restriction \nu = \pi \restriction \nu$, where $\nu = \nu(\dot{E}_\rho^W)$. Now $\mathrm{Ult}(R, \pi(\dot{E}_\rho^W))$ is not quite the last model of an iteration tree on K^c, since the ultrapowers do not come in the increasing length order. But this is easy to fix: we can find an embedding $\psi : \mathrm{Ult}(R, \pi(\dot{E}_\rho^W)) \to Q$, where Q comes from forming $\mathrm{Ult}(K^c, \pi(\dot{E}_\rho^W))$, and then doing the images of the ultrapowers in the iteration from K^c to R. We have $\psi \restriction \pi(\mu) = \mathrm{identity}$. We have then that the phalanx $\mathcal{D}$ with models $\langle R, Q, S\rangle$ and exchange ordinals $\langle \pi(\nu), \pi(\rho)\rangle$ is K^c-generated, and therefore is $\Omega + 1$ iterable. (cf. 6.9) We can use the maps $\langle \pi, \psi \circ \sigma, \pi\rangle$ to reduce trees on $\mathcal{B}$ to trees on $\mathcal{D}$, and hence $\mathcal{B}$ is $\Omega + 1$ iterable.

This completes the proof of Lemma 8.9, and hence of Theorem 8.8. □

Theorem 8.8 leads to the following characterization of K.

Theorem 8.10. *Suppose $K^c \models$ "There are no Woodin cardinals"; then K is the unique universal weasel which elementarily embeds into all universal weasels.*

Proof. Let M be a universal weasel. For $\alpha < \Omega$, we construct a linear iterate $\mathcal{P}_\alpha$ of M by hitting each total order zero measure above α once. (More precisely, let $\mathcal{U}_\nu$ be the νth total-on-M measure of order zero on the M-sequence which has critical point $\geq \alpha$. Let $\mathcal{T}$ be the linear iteration tree on M such that $E_\nu^{\mathcal{T}} = i_{0,\nu}^{\mathcal{T}}(\mathcal{U}_\nu)$ for all ν, and let $lh\ \mathcal{T} = \gamma+1 \leq \Omega+1$; then we set $P_\alpha = \mathcal{M}_\gamma^{\mathcal{T}}$.)

It is not hard to see that if $\alpha < \beta < \Omega$, then P_α is a linear iterate of P_β. For let $\mathcal{U}_\nu$ be the νth total measure of order zero on the M-sequence with critical point between α and β, and let $j : M \to P_\beta$ be the iteration map. We define a linear iteration tree $\mathcal{T}$ on P_β by: $E_\nu^\mathcal{T} = i_{0,\nu}^\mathcal{T}(j(\mathcal{U}_\nu))$. Letting $\gamma + 1 = lh\ \mathcal{T}$ (so that γ is the order type of $\{\nu \mid \mathcal{U}_\nu \text{ exists}\}$), we have $P_\alpha = \mathcal{M}_\gamma^\mathcal{T}$. (This comes down to the fact that if i and k are the embeddings associated to normal measures on distinct measurable cardinals, then $i(k) = k$.)

Since M is universal, $(\alpha^+)^M = \alpha^+$ for all but nonstationary many $\alpha \in A_0$, by 3.7 (1). The construction of P_β guarantees that no $\gamma \in A_0 - \beta$ is the critical point of a total-on-P_β extender from the P_β sequence. Therefore Ω is A_0-thick in P_β, for all $\beta < \Omega$. From 5.7 we then have that $K \cong \text{Def}(P_\beta)$, for all $\beta < \Omega$. Let $\pi_\beta : K \to P_\beta$ have range $\text{Def}(P_\beta)$.

Let $i_\beta : P_\beta \to P_0$ be the linear iteration map described above, so that $i_\beta'' \text{Def}(P_\beta) = \text{Def}(P_0)$ by 5.6. Since $i_\beta(\pi_\beta(\mu)) = \pi_0(\mu)$, we have $\pi_\beta(\mu) \leq \pi_0(\mu)$ for all β and μ. Now let us define $\sigma : K \to M$ as follows: for $x \in K$, pick any μ such that $x \in \mathcal{J}_\mu^K$, and let β be inaccessible and such that $\pi_0(\mu) < \beta$; then we set $\sigma(x) = \pi_\beta(x)$. Note here that $\pi_\beta(x) \in \mathcal{J}_{\pi_\beta(\mu)}^{P_\beta} \subseteq \mathcal{J}_\beta^{P_\beta} = \mathcal{J}_\beta^M$, so indeed $\sigma(x) \in M$. Also, if $\beta < \gamma$, then the iteration from P_γ to P_β has critical point $\geq \beta$, so that $Def(P_\gamma) \cap \mathcal{J}_\beta^{P_\gamma} = \text{Def}(P_\beta) \cap \mathcal{J}_\beta^{P_\beta}$. This implies that σ is well-defined. Finally, if $K \models \varphi[x]$, then $P_\beta \models \varphi[\pi_\beta(x)]$, so $M \models \varphi[\sigma(x)]$ because the iteration from M to P_β has critical point $\geq \beta$ and $\pi_\beta(x) \in \mathcal{J}_\beta^M$. This shows that σ is elementary.

Finally, we show uniqueness. Let $j : M \to K$ be elementary, where M is a universal weasel. Let $i : K \to M$ be elementary. Then $j \circ i : K \to K$ elementarily, so $j \circ i = \text{identity}$ by 8.8. It follows that $M = K$. □

We now consider the situation "below $0^{\mathbb{P}}$".

Definition 8.11. *A proper premouse $\mathcal{M}$ is below $0^{\mathbb{P}}$ iff whenever E is an extender on the $\mathcal{M}$-sequence, and $\kappa = crit(E)$, then $\mathcal{J}_\kappa^\mathcal{M} \models$ "There are no strong cardinals".*

One important way in which a premouse $\mathcal{M}$ below $0^{\mathbb{P}}$ is simple is that every normal iteration tree $\mathcal{T}$ on $\mathcal{M}$ is "almost linear". More precisely: if $\mathcal{T}$ is a normal tree on $\mathcal{M}$ and $T\text{-pred}(\beta+1) = \alpha$, then for some $n \in \omega$, $\beta = \alpha + n$ and $\text{crit}(E_\beta^\mathcal{T}) = \text{crit}(E_{\alpha+k}^\mathcal{T})$ for all $k \leq n$. Thus $\mathcal{T}$ misses being linear only in that it may hit the same critical point finitely many times in immediate succession (and thus branch finitely) before going on to critical points larger than the lengths of all preceding extenders. [Proof: It is enough to show that whenever $\alpha \leq \beta$ and $\text{crit}(E_\beta^\mathcal{T}) \leq \nu(E_\alpha^\mathcal{T})$, then $\text{crit}(E_\beta^\mathcal{T}) = \text{crit}(E_\alpha^\mathcal{T})$. So let $\alpha \leq \beta$, $\kappa = \text{crit}(E_\alpha^\mathcal{T})$, $\mu = \text{crit}(E_\beta^\mathcal{T})$, and suppose $\mu \leq \nu(E_\alpha^\mathcal{T})$. If $\kappa < \mu$, then there are arbitrarily large $\lambda < \mu$ such that $E_\alpha^\mathcal{T} \restriction \lambda$ is on the $\mathcal{M}_\beta^\mathcal{T}$ sequence, so $\mathcal{M}_\beta^\mathcal{T}$ is not below $0^{\mathbb{P}}$. If $\mu < \kappa$, then there are arbitrarily large $\lambda < \kappa$ such that $E^{\mathcal{T}}_{\beta} \restriction \lambda$ is on the $\mathcal{M}^{\mathcal{T}}_{\beta}$ sequence, so $\mathcal{M}^{\mathcal{T}}_{\beta}$ is not below $0^{\mathbb{P}}$. Both statements follow at once from the inital segment condition and the normality of $\mathcal{T}$.]

Lemma 8.12. *Suppose K^c is below $0^{\mathbb{P}}$. Let W and M be universal weasels, with Ω A_0-thick in M and W, and $\mu \subseteq Def(W)$. Let $i : W \to Q$ and $j : M \to Q$ come from coiteration, and $\pi : K \to M$ with ran $\pi = Def(M)$. Then $j \restriction sup(\pi''\mu) = identity$.*

Proof. Let $\mathcal{T}$ be the iteration tree on W producing $i = i^{\mathcal{T}}_{0\delta}$, and $\mathcal{U}$ the iteration tree on M producing j. Since we are below $0^{\mathbb{P}}$, $\mathcal{T}$ and $\mathcal{U}$ are almost linear. Suppose toward contradiction that $\text{crit}(j) < \pi(\gamma)$, where $\gamma < \mu$. Let $\kappa = \text{crit}(j) = \text{crit}(E^{\mathcal{U}}_0)$.

Since $i''\text{Def}(W) = \text{Def}(Q) = j''\text{Def}(M)$ by 5.6, we have $i(\gamma) = j(\pi(\gamma))$, and thus $\kappa < \sup(i''\mu)$. But now for any $\eta < \sup(i''\mu)$, either Q has the definability property at η or $\text{crit}(E^{\mathcal{T}}_{\alpha}) \leq \eta < \nu(E^{\mathcal{T}}_{\alpha})$ for some $\alpha T \delta$. (If $\eta < i(\xi)$ and the second disjunct fails, we can write $\eta = i(f)(\bar{a})$ where $\bar{a} \in [\eta]^{<\omega}$ and $f : \xi \to \xi$, so that $f \in \text{Def}(W)$ and hence $i(f) \in \text{Def}(Q)$.) Since $\kappa = \text{crit}(j)$, Q does not have the definability property at κ, and thus we can fix $\alpha+1 < lh\ \mathcal{T}$ such that $\text{crit}(E^{\mathcal{T}}_{\alpha}) \leq \kappa < \nu(E^{\mathcal{T}}_{\alpha})$.

We claim that $\kappa = \text{crit}(E^{\mathcal{T}}_{\alpha})$. This is where we make real use of the fact that our mice are below $0^{\mathbb{P}}$. For otherwise, letting $\theta = \text{crit}(E^{\mathcal{T}}_{\alpha})$, we have that $\mathcal{J}^M_{\kappa} \models \theta$ is a strong cardinal. This is because there are arbitrarily large $\lambda < \kappa$ such that $E^{\mathcal{T}}_{\alpha} \restriction \lambda$ is on the M-sequence. (Proof: κ is inaccessible in Q, and hence in $\mathcal{J}^{\mathcal{M}^{\mathcal{T}}_{\alpha}}_{\xi}$, where $\xi = lh\ E^{\mathcal{T}}_{\alpha}$. The initial segment condition gives arbitrarily large $\lambda < \kappa$ such that $E^{\mathcal{T}}_{\alpha} \restriction \lambda$ is on the $\mathcal{M}^{\mathcal{T}}_{\alpha}$ sequence. But $\mathcal{J}^M_{\kappa} = \mathcal{J}^{\mathcal{M}^{\mathcal{T}}_{\alpha}}_{\kappa}$.)

Let $i^{\mathcal{T}}_{\alpha\delta} : \mathcal{M}^{\mathcal{T}}_{\alpha} \to Q$ be the remainder of the iteration. Since $\mathcal{M}^{\mathcal{T}}_{\alpha}$ has the hull property at κ and $\kappa = \text{crit}(i^{\mathcal{T}}_{\alpha\delta})$, Q has the hull property at κ, and hence so does M. But then $i^{\mathcal{T}}_{\alpha\delta}(A) \cap \nu = j(A) \cap \nu$ for all $A \in P(\kappa)^M$, where $\nu = \inf(\nu(E^{\mathcal{T}}_{\alpha}), \nu(E^{\mathcal{U}}_0))$. It follows that $E^{\mathcal{T}}_{\alpha}$ is compatible with $E^{\mathcal{U}}_0$, the usual contradiction. □

Theorem 8.13. *Suppose K^c is below $0^{\mathbb{P}}$; then any universal weasel is a normal iterate of K, and any $j : K \to M$ is the iteration map associated to a normal iteration of K.*

Proof. The second statement actually implies the first, via 8.10, but we give an independent proof. Let $(\mathcal{T},\mathcal{U})$ be the coiteration of K with M. Since K^c is below $0^{\mathbb{P}}$, both $\mathcal{T}$ and $\mathcal{U}$ are normal, "almost linear" iteration trees. We wish to see $lh\ \mathcal{U} = 1$. If not, then $E^{\mathcal{U}}_0$ exists; let $\mu < \Omega$ be inaccessible and such that $lh\ E^{\mathcal{U}}_0 < \mu$. Let K^* and M^* come from K and M, respectively, by hitting each total order zero measure above μ. So Ω is A_0-thick in K^* and M^*, and $\mu \subseteq \text{Def}(K^*)$. If $(\mathcal{T}^*,\mathcal{U}^*)$ is the coiteration of K^* with M^*, then $E^{\mathcal{U}^*}_0 = E^{\mathcal{U}}_0$. This contradicts Lemma 8.12.

Next, let $j : K \to M$ be elementary. By the Dodd-Jensen lemma, M is universal, and hence there is a normal iteration tree $\mathcal{T}$ on K such that $M = \mathcal{M}_\gamma^\mathcal{T}$, for some $\gamma \leq \Omega$. We wish to show that $j = i_{0\gamma}^\mathcal{T}$. So fix $\eta < \Omega$; we shall show that $j(\eta) = i_{0\gamma}^\mathcal{T}(\eta)$.

Let $\beta < \Omega$ be such that $j(\eta) < \beta$ and $\forall\alpha\ (\text{crit}(E_\alpha^\mathcal{T}) < \beta \Rightarrow lh\ (E_\alpha^\mathcal{T}) < \beta)$. Let F be the length β extender derived from j, and let $j' : K \to M'$ be the canonical embedding of K into $\text{Ult}(K, F) = M'$. Clearly, M' agrees with M below β, and $j'(\eta) = j(\eta)$. Let $\mathcal{T}'$ be the normal iteration tree on K whose last model $\mathcal{M}_\delta^{\mathcal{T}'} = M'$. The agreement between M and M' implies that $\forall\alpha[(lh(E_\alpha^\mathcal{T}) < \beta \vee lh(E_\alpha^{\mathcal{T}'}) < \beta) \Rightarrow E_\alpha^\mathcal{T} = E_\alpha^{\mathcal{T}'}]$. It follows that if $i_{0\delta}^{\mathcal{T}'}(\eta) = j'(\eta)$, then $i_{0\gamma}^\mathcal{T}(\eta) = j(\eta)$, and we are done.

Let K^* come from K by hitting each total order zero measure with critical point above β once. Let M^* come from M' via the same process, using critical points above $j'(\beta)$. Clearly $j' : K^* \to M^*$, Ω is A_0-thick in K^* and M^*, and $\beta \subseteq \text{Def}(K^*)$. Also, $\{\alpha \mid j'(\alpha) = \alpha\}$ is A_0 thick in K^* and M^*, which is why we switched from j to j'.

Now let $i : K^* \to Q$ and $k : M^* \to Q$ be the iteration maps coming from the coiteration of K^* with M^*. Since K and K^* agree below β, as do M' and M^*, it will be enough to show that $i(\eta) = j'(\eta)$, for then $i(\eta) = i_{0\delta}^{\mathcal{T}'}(\eta)$ and we are done.But since j' has a thick class of fixed points, the proof of 5.6 gives $i''\ \text{Def}(K^*) = \text{Def}(Q) = (k \circ j')''\ \text{Def}(K^*)$. Thus $i(\eta) = k(j'(\eta))$. Since $k \restriction \beta = \text{identity}$ by 8.12, $i(\eta) = j'(\eta)$, as desired. □

We have developed the theory of K^c and K under the assumption that there is a measurable cardinal Ω, and so this is a tacit hypothesis in 8.13. The measurable cardinal is not needed for the theory of K "below $0^\mathbb{P}$", however. (See [DJKM].) Thus 8.13 does not require this tacit hypothesis.

We now sketch an argument which shows that the hypothesis of 8.13 that K^c is below $0^\mathbb{P}$ cannot be substantially weakened. The reason is that the conclusion of 8.13 implies, via work of Jensen and Mitchell, that $K \cap HC$ is Σ_5^1 in the codes. On the other hand, Woodin has shown that it is consistent that K^c is "below two strong cardinals", and yet $K \cap HC$ is not Σ_5^1 in the codes.

We sketch the definition of K "below $0^\mathbb{P}$" due to Mitchell and Jensen. Let us call α a closure point of a premouse $\mathcal{M}$ iff α is a limit of $\mathcal{M}$-cardinals and $\forall\beta < \alpha\exists\gamma < \alpha\forall\theta \in (\gamma, \alpha)$ ($\mathcal{J}_\theta^\mathcal{M}$ is active $\Rightarrow \text{crit}(E_\theta^\mathcal{M}) > \beta$). Let us call a premouse $\mathcal{M}$ α-good just in case $\mathcal{M}$ is iterable, $\rho_\omega(\mathcal{M}) = \alpha$, and either α is a closure point of $\mathcal{M}$ or there is a universal weasel W such that for some β: $\mathcal{M} = \mathcal{J}_\beta^W$, $\rho_\omega(\mathcal{J}_\gamma^W) \geq \alpha$ for all $\gamma \geq \beta$, and $\text{crit}(\dot{E}_\gamma^W) > \beta$ for all $\gamma > \beta$. Clearly, if W witnesses that $\mathcal{M}$ is α-good, then α is a cardinal of W. It can be shown that the relation R is Π_3^1, where $R(x, y) \Leftrightarrow (x, y \in {}^\omega\omega \wedge x$ codes a premouse $\mathcal{M} \wedge y$ codes an ordinal $\alpha \wedge \mathcal{M}$ is α-good). [If $\mathcal{M}, \alpha \in HC$, and α is not a closure point of $\mathcal{M}$, then $\mathcal{M}$ is α-good iff $\forall\mathcal{N} \in HC$ [$\exists\beta$ ($\mathcal{M} = \mathcal{J}_\beta^\mathcal{N} \wedge \forall\gamma > \beta(\text{crit}(\dot{E}_\gamma^\mathcal{N}) > \beta) \wedge$ "$\mathcal{N}$ is iterable via extenders with critical point $> \beta$") $\Rightarrow \mathcal{N}$ is iterable and $\rho_\omega(\mathcal{N}) \geq \alpha$.] This is Π^1_3 in the codes; the antecedent in the bracketed conditional is Π^1_2 because of the iterability assertion. The $\Rightarrow$ direction of this equivalence comes from the coiteration of a weasel W witnessing $\mathcal{M}$ is α-good with a potential counterexample $\mathcal{N}$ to the right hand side. For the $\Leftarrow$ direction, the weasel W which witnesses that $\mathcal{M}$ is α-good is "$K^c(\mathcal{M})$", the result of the construction of §1 modified so that it begins with $\mathcal{N}_0 = \mathcal{M}$ and only uses extenders with critical point $> \mathrm{OR}^{\mathcal{M}}$. We use here that α is not a closure point of $\mathcal{M}$. This, together with the other conditions, implies that $\beta+1$ is contained in every hull formed in the $K^c(\mathcal{M})$ construction, so that this construction does indeed produce the desired W. If α is a closure point, $\alpha \notin$ some hull is possible.]

Notice that if α is a cardinal of K, and $\rho_\omega(\mathcal{J}^K_\beta) = \alpha$, then $\mathcal{J}^K_\beta$ is α-good. This is clear if α is a closure point of K. If α is not a closure point of K, then the weasel W which witnesses that $\mathcal{J}^K_\beta$ is α-good is $\mathrm{Ult}(K, \dot{E}^K_\gamma)$, where γ is least such that $\gamma > \beta$ and $\mathrm{crit}(\dot{E}^K_\gamma) < \beta$, unless there is no such γ, in which case $W = K$ is the witness. (If there is such a γ, then $\mathrm{crit}(\dot{E}^K_\gamma) < \alpha$ for the least such γ. For if $\mathrm{crit}(\dot{E}^K_\gamma) = \alpha$, then α is a limit cardinal of K, so since α is not a closure point, there is a $\kappa < \alpha$ which is the critical point for extenders on the K-sequence with indices unbounded in α. The initial segment condition gives ν such that $\beta < \nu < \gamma$ and $\mathrm{crit}(\dot{E}^{\mathcal{P}}_\nu) = \kappa$, where $\mathcal{P} = \mathrm{Ult}(\mathcal{J}^K_\gamma, \dot{E}^K_\gamma)$ But $\dot{E}^{\mathcal{P}}_\nu = \dot{E}^K_\nu$ by coherence, and this contradicts the minimality of γ. Thus $\mathrm{crit}(\dot{E}^K_\gamma) < \alpha$, and therefore $\mathrm{Ult}(K, \dot{E}^K_\gamma)$ makes sense.)

Suppose that, conversely, whenever α is a cardinal of K, $\mathcal{J}^{\mathcal{M}}_\alpha = \mathcal{J}^K_\alpha$, and $\mathcal{M}$ is α-good, then $\mathcal{M} = \mathcal{J}^K_\beta$ for some β. It would follow that

$$\begin{aligned} \exists\gamma < (\alpha^+)^K(\mathcal{P} &= \mathcal{J}^K_\gamma) \\ \Leftrightarrow \exists\mathcal{M}(\mathcal{J}^{\mathcal{M}}_\alpha &= \mathcal{J}^K_\alpha \wedge \mathcal{M} \text{ is } \alpha\text{-good } \wedge \exists\gamma(\mathcal{P} = \mathcal{J}^{\mathcal{M}}_\gamma)). \end{aligned}$$

This easily implies that the function $\alpha \mapsto \mathcal{J}^K_\alpha$, restricted to HC, is Π^1_4 in the codes, and hence that $K \cap HC$ is Σ^1_5 in the codes. It is easy to see that if $\mathcal{M}$ is α-good, $\mathcal{J}^{\mathcal{M}}_\alpha = \mathcal{J}^K_\alpha$, and α is a closure point of $\mathcal{M}$, then $\mathcal{M} = \mathcal{J}^K_\beta$ for some β. This is because all critical points in the coiteration of $\mathcal{M}$ with K are $\geq \alpha$. So if $K \cap HC$ is not Σ^1_5 in the codes, there is a cardinal α of K which is not a closure point of K, and an α-good $\mathcal{M}$ such that $\mathcal{J}^{\mathcal{M}}_\alpha = \mathcal{J}^K_\alpha$ but $\mathcal{M}$ is not an initial segment of K.

Let $\mathcal{M}$ be α-good, as witnessed by W, and $\mathcal{J}^{\mathcal{M}}_\alpha = \mathcal{J}^K_\alpha$. If W is an iterate of K, then the iteration must use extenders of length $> \alpha$ because $\mathcal{J}^W_\alpha = \mathcal{J}^K_\alpha$, and this implies $\mathcal{M}$ is an initial segment of K because $\rho_\omega(\mathcal{M}) = \alpha$. So if $K \cap HC$ is not Σ^1_5 in the codes, then there is a universal weasel W which is not an iterate of K. By 8.10, there is nevertheless an elementary $j : K \to W$, and of course this j cannot be an iteration map.

We shall not attempt to sketch Woodin's proof that it is consistent that K^c has no Woodin cardinals (and is in fact "below two strong cardinals") and yet $K \cap HC$ is not Σ^1_5 in the codes. See [H].

Although we do not have the decisive characterization of embeddings of K given by 8.13, once we get past $0^{\mathbb{P}}$, we can prove the following consequence of the characterization.

Theorem 8.14. *Suppose $K^c \models$ There are no Woodin cardinals, and let M be $\Omega + 1$ iterable, $j : K \to M$ be elementary, and $\kappa = crit(j)$. Then*

(1) $P(\kappa)^M = P(\kappa)^K$,

(2) the trivial completion of the $(\kappa, \kappa + 1)$ extender derived from j is on the K-sequence,

(3) if $K \models$ "γ is regular but not measurable", then $j(\gamma) = \sup(j''\gamma)$.

Proof. We simply trace through the proofs of 8.12 and 8.13, and see what we get.

Let M, j, κ, and γ be as in the statement of 8.14. Let $\beta < \Omega$ be inaccessible with $\kappa, \gamma < \beta$. Let F be the length β extender derived from j. Let K^* be the witness that $\mathcal{J}^K_\beta$ is A_0-sound which is obtained from K by hitting each order zero measure above β. Let $M^* = \text{Ult}(K^*, F)$. Since M is universal by Dodd-Jensen, $(\alpha^+)^M = \alpha^+$ for all but nonstationary many $\alpha \in A_0$ by 3.7, and hence Ω is A_0 thick in M^*. Further, $\{\alpha \mid j^*(\alpha) = \alpha\}$ is A_0-thick in K^*, where $j^* : K^* \to M^*$ is the canonical embedding. It will be enough to show that $P(\kappa)^{M^*} = P(\kappa)^{K^*}$, that the $(\kappa, \kappa + 1)$ extender derived from j^* is on the K^* sequence, and that $j^*(\gamma) = \sup(j^{*''}\gamma)$.

Let $i : K^* \to Q$ and $k : M^* \to Q$ be the iteration maps coming from a coiteration of K^* with M^*. We have $\beta \subseteq \text{Def}(K^*)$, and since j^* has a thick class of fixed points, $\kappa \subseteq \text{Def}(M^*)$. Standard arguments then show that $i \restriction \kappa = (k \circ j^*) \restriction \kappa = \text{identity}$, and therefore $P(\kappa)^{K^*} = P(\kappa)^Q = P(\kappa)^{M^*}$. Further, since $\kappa = \text{crit}(k \circ j^*)$, $\kappa \notin \text{Def}(Q)$, and therefore $\kappa = \text{crit}(i)$. Since K^* and M^* have the hull property at κ, if $\text{crit}(k) = \kappa$ then the usual argument shows that the first extenders used along the branches of the iteration trees producing i and k are compatible, which is a contradiction. Therefore $\text{crit}(k) > \kappa$. But then the $(\kappa, \kappa + 1)$ extender derived from i is the same as the $(\kappa, \kappa + 1)$ extender derived from j^*. Since i is an iteration map, the $(\kappa, \kappa + 1)$ extender derived from i is on the K^*-sequence, and we have proved (1) and (2).

Since $K^* \models$ "γ is regular but not measurable", and i is an iteration map, $i(\gamma) = \sup(i''\gamma)$. Since $\beta \subseteq \text{Def}(K^*)$, $i \restriction \beta = (k \circ j^*) \restriction \beta$, and therefore $(k \circ j^*)(\gamma) = \sup((k \circ j^*)''\gamma)$. It follows that $j^*(\gamma) = \sup(j^{*''}\gamma)$, which proves (3). □

Clearly, 8.14 can be pushed a little further, in that more of j is given by an iteration of K than its normal measure part.

We conclude this section by giving another proof that if there is a strongly compact cardinal, then there is an inner model with a Woodin cardinal. The proof we gave in §7 required an excursion into descriptive set theory, whereas this proof does not.

Lemma 8.15. *Suppose $K^c \models$ There are no Woodin cardinals, and let $\mu < \Omega$ be measurable; then $(\mu^+)^K = \mu^+$.*

Proof. Let $j : V \to M$ with $\mathrm{crit}(j) = \mu$. Since μ is measurable, we can prove all the results of §1 - §6 with μ replacing Ω. (Let K^c_μ be K^c as constructed in V_μ. If $K^c_\mu \models$ There is a Woodin cardinal, then there is a model of height Ω having a Woodin cardinal $\delta < \Omega$. As we remarked in §2, this is impossible if $K^c \models$ There are no Woodin cardinals.) Let K_μ be the model constructed in §5, but "below μ". Let $\mathcal{U} = \{X \subseteq \mu \mid \mu \in j(X)\}$ be the normal ultrafilter generated by j. By 5.18, for $\mathcal{U}$ a.e. $\alpha < \mu$, $(\alpha^+)^{K_\mu} = \alpha^+$. But now the results of §6 give an inductive definition of K_μ which is precisely the same as that of $\mathcal{J}^K_\mu$, and therefore $K_\mu = \mathcal{J}^K_\mu$. It follows that $(\alpha^+)^K = \alpha^+$ for $\mathcal{U}$ a.e. $\alpha < \mu$.

But then $(\mu^+)^{K^M} = (\mu^+)^M = \mu^+$. Since $j : K \to K^M$, we have $(\mu^+)^K = (\mu^+)^{K^M}$ by 8.14. So $(\mu^+)^K = \mu^+$, as desired. □

Theorem 8.16. *Let Ω be measurable, and let $\mu < \Omega$ be μ^+-strongly compact; then $K^c \models$ There is a Woodin cardinal.*

Proof. Suppose otherwise. Let $j : V \to M$ come from the ultrapower of V by a fine, μ-complete ultrafilter on $P_\mu(\mu^+)$. It is well known that $j(\mu^+) > \sup(j''\mu^+)$. But $\mu^+ = (\mu^+)^K$ by 8.15, and j is continuous at $(\mu^+)^K$ by (3) of 8.14. This is a contradiction. □

§9. A general iterability theorem

In this section we give a full proof of the iterability facts we have used. The proof results from an amalgamation of §12 of [FSIT], §4 of [IT], and §2 of this paper. Given a premouse $\mathcal{M}$ of the construction $\mathbb{C}$ of [FSIT], and an iteration tree $\mathcal{T}$ on $\mathcal{M}$, §12 of [FSIT] shows how to use the background extenders of $\mathbb{C}$ to "enlarge $\mathcal{T}$" to an iteration tree $\mathcal{U}$ on V. The good behavior of $\mathcal{U}$ guarantees that of $\mathcal{T}$. That $\mathcal{U}$ is well-behaved is shown in §4 of [IT], by realizing in V the models $\mathcal{M}_\alpha^{\bar{\mathcal{U}}}$ occurring on countable elementary submodels $\bar{\mathcal{U}}$ of $\mathcal{U}$. However, since the construction $\mathbb{C}$ of the present paper does not involve background extenders over V, we cannot in the current situation enlarge $\mathcal{T}$ to a tree on V. Instead, we shall run the enlargement process of [FSIT] and the realization process of [IT] simultaneously, making do with the partial background extenders of $\mathbb{C}$ as in §2.

We have also re-organized and streamlined the construction of §4 of [IT]. Moreover, in order to cover all our applications, we shall consider more than just iteration trees on premice.

Definition 9.1. *A creature is a structure which is either a premouse, a psuedo-premouse, or a bicephalus.*

Let $\mathbb{C}$ be the construction of §1, that is,

$$\mathbb{C} = \langle \mathcal{N}_\xi \mid \xi < \Omega \wedge \mathcal{N}_\xi \text{ is defined} \rangle .$$

Definition 9.2. *$\mathcal{M}$ is a creature of $\mathbb{C}$ just in case for some j, ξ*

(a) $\mathcal{M} = \mathfrak{C}_j(\mathcal{N}_\xi)$, or

(b) $\mathcal{M} = (\mathfrak{C}_\omega(\mathcal{N}_\xi), F)$, $\mathcal{M}$ is a psuedo-premouse, and letting $\kappa = crit(F)$, $\forall \mathcal{A} \subseteq P(\kappa)^{\mathcal{M}} (|\mathcal{A}| \leq \omega \Rightarrow F$ has a certificate on $\mathcal{A})$, or

(c) $\mathcal{M} = (\mathfrak{C}_\omega(\mathcal{N}_\xi), F_0, F_1)$, $\mathcal{M}$ is a bicephalus, and letting $i \in \{0,1\}$ and $\kappa_i = crit(F_i)$, $\forall \mathcal{A} \subseteq P(\kappa_i)^{\mathcal{M}}$ $(|\mathcal{A}| \leq \omega \Rightarrow F_i$ has a certificate on $\mathcal{A})$.

We say $\mathcal{M}$ *is $\mathbb{C}$-exotic* just in case condition (a) above fails to hold.

If $\mathcal{M}$ is a creature of $\mathbb{C}$ which is not a premouse, then $\mathcal{M}$ must be $\mathbb{C}$-exotic, but we do not know whether the converse is true. If $\mathcal{M} = \mathfrak{C}_j(\mathcal{N}_\xi)$, then we call (j, ξ) an index (in $\mathbb{C}$) for $\mathcal{M}$; a non-exotic $\mathcal{M}$ can have more than one such index, but all its indices have the same second coordinate. If $\mathcal{M}$ is $\mathbb{C}$-exotic, it must be of the form $(\mathfrak{C}_\omega(\mathcal{N}_\xi), F)$ or $(\mathfrak{C}_\omega(\mathcal{N}_\xi), F_0, F_1)$, and then we say $(0, \xi)$ is an index (in $\mathbb{C}$) for $\mathcal{M}$. A $\mathbb{C}$-exotic creature of $\mathbb{C}$ has exactly one index in $\mathbb{C}$. By ind($\mathcal{M}$) we mean the common second coordinate of all indices of $\mathcal{M}$.

Recall that a coarse premouse is a structure $\mathcal{M} = (M, \in, \delta)$ such that M is transitive, power admissible, satisfies choice, infinity, and the full separation schema, satisfies the full collection schema for domains contained in V_δ, and such that $\omega\delta = \delta$ and ${}^\omega M \subseteq M$.

Definition 9.3. *If $\mathcal{M}$ is a coarse premouse, then $\mathbb{C}^{\mathcal{M}} = \langle \mathcal{N}_\xi^{\mathcal{M}} \mid \xi < \delta^{\mathcal{M}}$ and $\mathcal{N}_\xi^{\mathcal{M}}$ exists$\rangle$ is the construction of §1 as done inside $\mathcal{M}$, up to stage $\delta^{\mathcal{M}}$.*

Thus $\mathbb{C} = \mathbb{C}^{\mathcal{M}}$ for all coarse premice $\mathcal{M}$ such that $\delta^{\mathcal{M}} = \Omega$ and $V_\Omega^{\mathcal{M}} = V_\Omega$. Notice that for any coarse premouse $\mathcal{M}$, $\mathcal{N}_\xi^{\mathcal{M}} \in V_{\delta^{\mathcal{M}}}^{\mathcal{M}}$ whenever $\mathcal{N}_\xi^{\mathcal{M}}$ exists. Further , there are in $V_{\delta^{\mathcal{M}}}$ certificates for all extenders put into models of $\mathbb{C}^{\mathcal{M}}$.
By convention, all creatures are 0-sound. The notion of a weak 0-embedding extends in an obvious way to creatures which are not premice. If $\pi : \mathcal{M} \to \mathcal{N}$ and $\mathcal{M}$ and $\mathcal{N}$ are psuedo-premice, then π is a weak 0-embedding just in case π is $r\Sigma_0$ elementary, and for some cofinal $X \subseteq \mathrm{OR}^{\mathcal{M}}$, π is $r\Sigma_1$ elementary on parameters from X. If $\pi : \mathcal{M} \to \mathcal{N}$ where $\mathcal{M} = (\mathcal{M}', F_0, F_1)$ and $\mathcal{N} = (\mathcal{N}', G_0, G_1)$ are bicephali, then π is a weak 0-embedding just in case it is a weak 0-embedding from $(\mathcal{M}', F_0)$ to $(\mathcal{N}', G_0)$ and a weak 0-embedding from $(\mathcal{M}', F_1)$ to $(\mathcal{N}', G_1)$. For $k > 0$, we shall consider k-soundness and weak k-embeddings only as applied to premice.

Definition 9.4. *Let $\mathcal{M}$ be a creature and let $k \leq \omega$; then $(\mathcal{R}, Q, \pi)$ is a k-realization of $\mathcal{M}$ just in case $\mathcal{R}$ is a coarse premouse and*

(a) *Q is a creature of $\mathbb{C}^{\mathcal{R}}$ of the same type as $\mathcal{M}$, and if $k > 0$ then $\mathcal{M}$ is premouse and $Q = \mathfrak{C}_k(\mathcal{N}_\xi)^{\mathcal{R}}$ for some ξ,*

(b) *π is a weak k-embedding from $\mathcal{M}$ into Q, and*

(c) *$\pi, \mathcal{M} \in \mathcal{R}$.*

In the situation of 9.4, if $\mathcal{M}$ is a premouse then the ordinal ξ as in (a) is determined uniquely by Q and k.

If $\mathcal{M}$ is a creature and $\omega\beta = \mathrm{OR}^{\mathcal{M}}$, then we set $\mathcal{J}_\beta^{\mathcal{M}} = \mathcal{M}$. If $\omega\beta < \mathrm{OR}^{\mathcal{M}}$, then we let $\mathcal{J}_\beta^{\mathcal{M}}$ be the unique premouse Q such that Q is an initial segment of $\mathcal{M}$ and $\omega\beta = \mathrm{OR}^Q$.

Definition 9.5. *Two creatures $\mathcal{M}$ and $\mathcal{N}$ agree below γ just in case for all $\beta < \gamma$, $\mathcal{J}_\beta^{\mathcal{M}} = \mathcal{J}_\beta^{\mathcal{N}}$.*

We wish to consider iteration trees whose base is a family of creatures.

Definition 9.6. *A phalanx of creatures is a pair*

$$(\langle(\mathcal{M}_\beta, k_\beta) \mid \beta \leq \alpha\rangle, \langle(\nu_\beta, \lambda_\beta) \mid \beta < \alpha\rangle)$$

such that for all $\beta \leq \alpha$

(1) *$\mathcal{M}_\beta$ is a creature, $k_\beta \leq \omega$, and if $k_\beta \neq 0$ then $\mathcal{M}_\beta$ is a k_β-sound premouse;*

(2) *if $\beta < \gamma < \alpha$, then $\nu_\beta < \nu_\gamma$;*

(3) *if $\beta < \gamma \leq \alpha$, then λ_β is the least $\eta \geq \nu_\beta$ such that $\mathcal{M}_\gamma \models \eta$ is a cardinal, and moreover, $\rho_{k_\gamma}(\mathcal{M}_\gamma) > \lambda_\beta$;*

(4) *$\lambda_\beta \leq OR^{\mathcal{M}_\beta}$; and*

(5) *if* $\beta < \gamma \leq \alpha$, *then* $\mathcal{M}_\beta$ *agrees with* $\mathcal{M}_\gamma$ *below* λ_β.

If $\mathcal{B}$ is a phalanx of creatures, say $\mathcal{B} = (\langle(\mathcal{M}_\beta, k_\beta) \mid \beta \leq \alpha\rangle, \langle(\nu_\beta, \lambda_\beta) \mid \beta < \alpha\rangle)$, then we set $\mathcal{M}^{\mathcal{B}}_\beta = \mathcal{M}_\beta$, $\deg^{\mathcal{B}}(\beta) = k_\beta$, $\nu(\beta, \mathcal{B}) = \nu_\beta$ and $\lambda(\beta, \mathcal{B}) = \lambda_\beta$. We also set $lh(\mathcal{B}) = \alpha + 1$. Notice that, because of (3), the $\lambda(\beta, \mathcal{B})$'s are determined by the $\nu(\beta, \mathcal{B})$'s and the $\mathcal{M}^{\mathcal{B}}_\beta$'s.

If $\mathcal{B}$ is a simple phalanx in the sense of §6, then $\mathcal{B}$ becomes a phalanx in the sense of 9.6 if we set $k_\beta = \omega$ for all $\beta < lh(\mathcal{B})$, and $\nu(\beta, \mathcal{B}) = \lambda(\beta, \mathcal{B})$ for all $\beta + 1 < lh(\mathcal{B})$. The notion of an iteration tree on a simple phalanx, as defined in §6, extends in an obvious way to phalanxes as defined in 9.6.

Definition 9.7. *Let* $\mathcal{T}$ *be an iteration tree of length* $\theta + 1$ *on a phalanx* $\mathcal{B}$ *of length* $\alpha + 1$. *Then:*

(i) (a) *for* $\beta \leq \alpha$, $\deg^{\mathcal{T}}(\beta) = \deg^{\mathcal{B}}(\beta)$, *and for* $\alpha < \beta \leq \theta$, $\deg^{\mathcal{T}}(\beta)$ *is the unique* $k \leq \omega$ *such that* $\mathcal{M}^{\mathcal{T}}_\beta = Ult_k((\mathcal{M}^{\mathcal{T}}_\beta)^*, E^{\mathcal{T}}_\beta)$ *if* β *is a successor, and the eventual value of* $\deg^{\mathcal{T}}(\gamma)$ *for* $\gamma T \beta$ *sufficiently large if* β *is a limit;*

(b) *for* $\beta < \alpha$, $\nu(\beta, \mathcal{T}) = \nu(\beta, \mathcal{B})$ *and* $\lambda(\beta, \mathcal{T}) = \lambda(\beta, \mathcal{B})$, *while for* $\alpha \leq \beta < \theta$, $\nu(\beta, \mathcal{T}) = \nu(E^{\mathcal{T}}_\beta)$ *and* $\lambda(\beta, \mathcal{T}) = lh(E^{\mathcal{T}}_\beta)$.

(ii) $\Phi(\mathcal{T})$ *is the unique phalanx* $\mathcal{D}$ *such that* $lh(\mathcal{D}$ *such that* $lh(\mathcal{D}) = \theta + 1$ *and*

(a) $\mathcal{M}^{\mathcal{D}}_\beta = \mathcal{M}^{\mathcal{T}}_\beta$ *and* $\deg^{\mathcal{D}}(\beta) = \deg^{\mathcal{T}}(\beta)$ *for all* $\beta \leq \theta$,

(b) $\nu(\beta, \mathcal{D}) = \nu(\beta, \mathcal{T})$ *and* $\lambda(\beta, \mathcal{D}) = \lambda(\beta, \mathcal{T})$ *for all* $\beta < \theta$.

A realization of a phalanx $\mathcal{B}$ will be a family of realizations of the creatures occurring in $\mathcal{B}$. We shall demand that these realizations agree with one another in a certain way. In order to explain this agreement condition, we now recall the terminology associated to "resurrection" in §12 of [FSIT].

Let $\mathcal{M}$ be a creature $\omega\alpha = \mathrm{OR}^{\mathcal{M}}$, and $t \leq \omega$. Suppose $t = 0$ if $\mathcal{M}$ is not a premouse. Let $\omega\lambda \leq \mathrm{OR}^{\mathcal{M}}$. Set

$$\langle \beta_0, k_0 \rangle = \langle \lambda, 0 \rangle,$$

and

$$\begin{aligned} \langle \beta_{i+1}, k_{i+1} \rangle = {} & \text{lexicographically least pair } \langle \beta, k \rangle. \text{ such that} \\ & \langle \lambda, 0 \rangle \leq \langle \beta_i, k_i \rangle \leq_{lex} \langle \beta, k \rangle \leq_{\text{lex}} \langle \alpha, t \rangle \\ & \text{and } \rho_k(\mathcal{J}^{\mathcal{M}}_\beta) < \rho_{k_i}(\mathcal{J}^{\mathcal{M}}_{\beta_i}), \end{aligned}$$

where $\langle \beta_{i+1}, k_{i+1} \rangle$ is undefined if no such pair exists. Let i be largest such that $\langle \beta_i, k_i \rangle$ is defined; then we call $\langle \langle \beta_0, k_0 \rangle, \ldots, \langle \beta_i, k_i \rangle \rangle$ the (t, λ) *dropdown sequence of* $\mathcal{M}$. It is clear that if $\langle \langle \beta_e, k_e \rangle \mid e \leq i \rangle$ is the (t, λ) dropdown sequence of $\mathcal{M}$, then $\langle \beta_e, k_e \rangle <_{\text{lex}} \langle \beta_{e+1}, k_{e+1} \rangle$ for all $e < i$, and $0 < k_e < \omega$ for all $e \leq i$ such that $e > 0$. Also, letting $\omega\alpha = \mathrm{OR}^{\mathcal{M}}$,

$$\{\rho_k(\mathcal{J}^{\mathcal{M}}_\beta) \mid \lambda \leq \beta \wedge (\beta, k) \leq_{\text{lex}} (\alpha, t) \wedge \rho_k(\mathcal{J}^{\mathcal{M}}_\beta) \leq \lambda\} = \{\rho_{k_e}(\mathcal{J}^{\mathcal{M}}_{\beta_e}) \mid e \leq i\}.$$

The following lemma is proved in §12 of [FSIT]. We gave its proof in a typical special case in Lemma 2.6 of these notes.

Lemma 9.8. *Let $\mathcal{M}$ be a creature of $\mathbb{C}$ with index (t,ξ), let $\langle\langle\beta_e,k_e\rangle \mid e\le i\rangle$ be the (t,λ) dropdown sequence of $\mathcal{M}$. Then there is a unique $\gamma\le\xi$ such that $\mathcal{J}^{\mathcal{M}}_{\beta_i}$ is a creature of $\mathbb{C}$ with index (k_i,γ).*

Now let $\mathcal{M}$ be a creature of $\mathbb{C}$ with index (t,ξ), and let $\omega\lambda\le\mathrm{OR}^{\mathcal{M}}$. We define the $(\mathcal{M},t,\xi)$ *resurrection sequence for* λ as follows. Let $\langle\beta,k\rangle$ be the last term in the (t,λ) dropdown sequence of $\mathcal{M}$. If $\langle\beta,k\rangle=\langle\lambda,0\rangle$, then the $(\mathcal{M},t,\xi)$ resurrection sequence for λ is empty. If $\langle\beta,k\rangle\neq\langle\lambda,0\rangle$ (so that $k>0$), then let $\gamma\le\xi$ be unique so that $\mathcal{J}^{\mathcal{M}}_{\beta}=\mathfrak{C}_k(\mathcal{N}_\gamma)$, as given by 9.8. Let

$$\pi:\mathfrak{C}_k(\mathcal{N}_\gamma)\to\mathfrak{C}_{k-1}(\mathcal{N}_\gamma)$$

be the canonical embedding. Then the $(\mathcal{M},t,\xi)$ resurrection sequence for λ is $\langle\beta,k,\gamma,\pi\rangle^\frown s$, where s is the $(\mathfrak{C}_{k-1}(\mathcal{N}_\gamma),k-1,\gamma)$ resurrection sequence for $\pi(\lambda)$. Here, as usual, if $\lambda=\mathrm{OR}\cap\mathfrak{C}_k(\mathcal{N}_\gamma)$, then $\pi(\lambda)=\mathrm{OR}\cap\mathfrak{C}_{k-1}(\mathcal{N}_\gamma)$ by convention. Notice $(\gamma,k-1)<_{\mathrm{lex}}(\xi,t)$, so this is indeed a legitimate inductive definition.

Now suppose $\mathcal{M}$ is a creature of $\mathbb{C}$ with index (t,ξ), $\omega\lambda\le\mathrm{OR}^{\mathcal{M}}$, $\langle\langle\beta_e,k_e\rangle \mid e\le i\rangle$ is the (t,λ) dropdown sequence of $\mathcal{M}$, and $\langle\langle\delta_e,\ell_e,\gamma_e,\pi_e\rangle \mid e\le s\rangle$ is the $(\mathcal{M},t,\xi)$ resurrection sequence for λ. As explained in §12 of [FSIT], we can find stages

$$i\le e_1<e_2<\cdots<e_{i-1}=s$$

such that for $1\le j\le i-1$,

$$\langle\delta_{e_j},\ell_{e_j}\rangle=\pi_{e_j-1}\circ\pi_{e_j-2}\circ\cdots\circ\pi_0(\langle\beta_{i-j},k_{i-j}\rangle)\,.$$

We set $e_0=0$, and interpret "$\pi_{e_0-1}\circ\cdots\circ\pi_0$" as standing for the identity embedding; this makes the equation just displayed true for $j=0$ as well. Set

$$\sigma_{i-j}=\pi_{e_j}\circ\pi_{e_j-1}\circ\cdots\circ\pi_0$$

so that

$$\sigma_{i-j}:\mathcal{J}^{\mathcal{M}}_{\beta_{i-j}}\to\mathfrak{C}_{\ell_{e_j}-1}(\mathcal{N}_{\gamma_{e_j}})$$

is an $\ell_{e_j}-1$ embedding, for $0\le j\le i-1$. In order to simplify the indexing a bit, we set $\tau_{i-j}=\gamma_{e_j}$ for $0\le j\le i-1$. Notice that $k_{i-j}=\ell_{e_j}$. Thus, setting $p=i-j$, we have that for $1\le p\le i$,

$$\sigma_p:\mathcal{J}^{\mathcal{M}}_{\beta_p}\to\mathfrak{C}_{k_p-1}(\mathcal{N}_{\tau_p})$$

is a k_p-1 embedding. Let us set $\mathrm{Res}_p=\mathfrak{C}_{k_p-1}(\mathcal{N}_{\tau_p})$.

Definition 9.9. *In the situation described above, we call (σ_p,Res_p) the pth partial resurrection of λ from stage (t,ξ).*

The partial resurrections of λ from stage (t,ξ) agree with one another in the following way. For $1 \le p \le i$, let

$$\kappa_p = \rho_{k_p}(\mathcal{J}^{\mathcal{M}}_{\beta_p}).$$

Then one can check without too much difficulty that $\kappa_1 > \kappa_2 > \cdots > \kappa_i$, and

$$p < q \Rightarrow \sigma_p \restriction \kappa_{q-1} = \sigma_q \restriction \kappa_{q-1},$$

and

$$p < q \Rightarrow \text{Res}_p \text{ and } \text{Res}_q \text{ agree below } \sup(\sigma_q''\kappa_{q-1}).$$

Definition 9.10. *In the situation described above, we call (σ, Res) the complete resurrection of λ from $(\mathcal{M},t,\xi)$ if and only if*

(a) *the $(\mathcal{M},t,\xi)$ resurrection sequence for λ is empty, and $(\sigma, Res) = (identity, \mathcal{M})$, or*

(b) *the $(\mathcal{M},t,\xi)$ resurrection sequence for λ is nonempty, and $(\sigma, Res) = (\sigma_1, Res_1)$.*

Notice that in either case of 9.10, Res is a creature of $\mathbb{C}$ with index (k,γ), for some $(\gamma,k) \le_{\text{lex}} (\xi,t)$. If Res is $\mathbb{C}$-exotic, then 9.10 (a) must hold.

Of course, the notions associated to resurrection can be interpreted in any coarse premouse $\mathcal{R}$, using $\mathbb{C}^{\mathcal{R}}$, and not just in V. We shall do this in the following.

Let (σ, Res) be the complete resurrection of λ from $(\mathcal{M},t,\xi)$. Suppose $\mathcal{J}^{\mathcal{M}}_{\lambda}$ is active, which is a case of particular interest. If $(t,\xi) = (0,\lambda)$, then $\mathcal{M} = \mathcal{J}^{\mathcal{M}}_{\lambda} = \text{Res}$, and σ is the identity. Otherwise, $\langle\lambda,1\rangle \le_{\text{lex}} \langle\xi,t\rangle$, so $\langle\beta_1,k_1\rangle = \langle\lambda,1\rangle$. It follows that $\text{Res} = \mathcal{N}_\gamma$ for some $\gamma \le \xi$, and $\sigma : \mathcal{J}^{\mathcal{M}}_{\lambda} \to \mathcal{N}_\gamma$ is a 0-embedding.

Definition 9.11. *Let $\mathcal{B}$ be a phalanx of length $\alpha+1$. Then a realization of $\mathcal{B}$ is a sequence $\langle(\mathcal{R}_\beta, Q_\beta, \pi_\beta) \mid \beta \le \alpha\rangle$ such that*

(1) *for all $\beta \le \alpha$, $(\mathcal{R}_\beta, Q_\beta, \pi_\beta)$ is a $\deg^{\mathcal{B}}(\beta)$-realization of $\mathcal{M}^{\mathcal{B}}_\beta$, and*

(2) *if $\beta < \gamma \le \alpha$, and τ is the unique ordinal ξ such that $(\deg^{\mathcal{B}}(\beta),\xi)$ is an index of Q_β in $\mathbb{C}^{\mathcal{R}_\beta}$, and $\lambda_\beta = \lambda(\beta,\mathcal{B})$, and $(\sigma^\beta, Res^\beta)$ is the complete resurrection of $\pi_\beta(\lambda_\beta)$ from $(Q_\beta, \deg^{\mathcal{B}}(\beta), \tau)$, and $\nu_\beta = \nu(\beta,\mathcal{B})$, then*

(a) *$V^{\mathcal{R}_\beta}_\mu = V^{\mathcal{R}_\gamma}_\mu$, and $V^{\mathcal{R}_\beta}_{\mu+1} \subseteq V^{\mathcal{R}_\gamma}_{\mu+1}$, for $\mu = \sigma^\beta \circ \pi_\beta(\nu_\beta)$,*

(b) *Res^β agrees with Q_γ below $\sigma^\beta \circ \pi_\beta(\lambda_\beta)$,*

(c) *$(\sigma^\beta \circ \pi_\beta) \restriction \lambda_\beta = \pi_\gamma \restriction \lambda_\beta$, and*

(d) *$(\sigma^\beta \circ \pi_\beta)(\lambda_\beta) \le \pi_\gamma(\lambda_\beta)$.*

If $\mathcal{E} = \langle(\mathcal{R}_\beta, Q_\beta, \pi_\beta) \mid \beta \le \alpha\rangle$ is a realization of $\mathcal{B}$, then we write $\mathcal{R}^{\mathcal{E}}_\beta$ for $\mathcal{R}_\beta$, etc.

Definition 9.12. *Let $\mathcal{B}$ be a phalanx of length $\alpha+1$, $\mathcal{E}$ a realization of $\mathcal{B}$, and $\mathcal{T}$ a putative iteration tree on $\mathcal{B}$. Let $\alpha+1 \le \gamma < lh\ \mathcal{T}$, and let $\beta \le \alpha+1$ and $\beta T \gamma$. We call a pair $(\mathcal{P},\sigma)$ an $\mathcal{E}$-realization of $\mathcal{M}^{\mathcal{T}}_\gamma$ if and only if*

(1) $(\mathcal{R}^{\mathcal{E}}_\beta, \mathcal{P}, \sigma)$ *is a* $\deg^{\mathcal{T}}(\gamma)$ *realization of* $\mathcal{M}^{\mathcal{T}}_\gamma$,

(2) *if* Q_β *has index* $(\deg^{\mathcal{B}}(\beta), \xi)$ *and* $\mathcal{P}$ *has index* $(\deg^{\mathcal{T}}(\gamma), \theta)$ *in the construction of* $\mathcal{R}^{\mathcal{E}}_\beta$, *then* $\theta \leq \xi$, *and*

$$D^{\mathcal{T}} \cap [\beta, \gamma]_T \neq \phi \Leftrightarrow \theta < \xi\,,$$

(3) *if* $D^{\mathcal{T}} \cap [\beta, \gamma]_T = \phi$ *and* $\deg^{\mathcal{T}}(\gamma) = \deg^{\mathcal{B}}(\beta)$, *then* $\mathcal{P} = Q_\beta$ *and* $\pi^{\mathcal{E}}_\beta = \sigma \circ i^{\mathcal{T}}_{\beta,\gamma}$.

Definition 9.13. *Let* $\mathcal{T}$ *be an iteration tree on* $\mathcal{B}$, *and* b *a maximal branch of* $\mathcal{T}$ *such that* $D^{\mathcal{T}} \cap b$ *is finite. Then an* $\mathcal{E}$*-realization of* b *is just an* $\mathcal{E}$*-realization of* $\mathcal{M}^{\mathcal{S}}_\gamma$, *where* $\gamma = \sup b$ *and* $\mathcal{S}$ *is the putative iteration tree of length* $\gamma + 1$ *such that* $\mathcal{S} \restriction \gamma = \mathcal{T} \restriction \gamma$ *and* $b = \{\eta \mid \eta S \gamma\}$. *We say* b *is* $\mathcal{E}$*-realizable iff there is an* $\mathcal{E}$*-realization of* b.

We can now state the main result of this section. Recall that a cutoff point of a coarse premouse $(M, \in, \delta)$ is an ordinal $\theta \in M$ such that $(V^M_\theta, \in, \delta)$ is a coarse premouse. We say that $\mathcal{M}$ has α cutoff points if the order type of the set of cutoff points of $\mathcal{M}$ is at least α.

Theorem 9.14. *Let* $\mathcal{B}$ *be a hereditarily countable phalanx, and let* $\mathcal{E}$ *be a realization of* $\mathcal{B}$ *such that* $\forall \alpha < lh(\mathcal{B})$ ($\mathcal{R}^{\mathcal{E}}_\alpha$ *has* $\delta^{\mathcal{R}^{\mathcal{E}}_\alpha}$ *cutoff points). Let* $\mathcal{T}$ *be a countable putative normal iteration tree on* $\mathcal{B}$. *Then either*

(1) $\mathcal{T}$ *has a maximal,* $\mathcal{E}$*-realizable branch, or*

(2) $\mathcal{T}$ *has a last model* $\mathcal{M}^{\mathcal{T}}_\gamma$, *and this model is* $\mathcal{E}$*-realizable.*

Proof. Fix $\mathcal{E}_0$, a realization of $\mathcal{B}_0$ as in the hypotheses, and $\mathcal{T}$ a putative normal iteration tree of countable length θ on $\mathcal{B}_0$. We shall consider no iteration trees but $\mathcal{T}$ in the proof to follow, and so we set $\mathcal{M}_\beta = \mathcal{M}^{\mathcal{T}}_\beta$, $E_\beta = E^{\mathcal{T}}_\beta$, $\nu_\beta = \nu(\beta, \mathcal{T})$, $\lambda_\beta = \lambda(\beta, \mathcal{T})$, and $\deg(\beta) = \deg^{\mathcal{T}}(\beta)$. Let $n^* : \theta \longrightarrow \omega$ be one-one, and set

$$n(\alpha) = \inf\{n^*(\beta) \mid \alpha = \beta \quad \text{or} \quad \alpha T \beta\}\,.$$

Clearly $\alpha T \beta \Rightarrow n(\alpha) \leq n(\beta)$, and for λ a limit $< \theta$, $n(\lambda)$ is the eventual value of $n(\beta)$ for all sufficiently large $\beta T \lambda$. Notice that if $n(\alpha) = n(\beta)$, then $\alpha T \beta$ or $\beta T \alpha$ or $\alpha = \beta$. Also, for b a branch of T,

$$b \text{ is maximal } \Leftrightarrow \sup\{n(\alpha) \mid \alpha \in b\} = \omega\,.$$

For $\alpha, \beta < \theta$, we say

$$\alpha \text{ survives at } \beta \quad \Leftrightarrow [\alpha = \beta \vee (\alpha T \beta \wedge n(\alpha) = n(\beta) \wedge \forall \gamma(\alpha < \gamma < \beta \wedge \gamma \notin (\alpha, \beta)_T) \Rightarrow n(\alpha) < n(\gamma)))]\,.$$

It is easy to see that if α survives at β and β survives at γ, then α survives at γ. Also, if α survives at γ and $\alpha T \beta T \gamma$, then α survives at β and β survives at γ. One can also easily see that if λ is a limit, then all sufficiently large $\beta T \lambda$ survive at λ, and that for b a branch of T,

$$b \text{ is maximal } \Leftrightarrow \forall \alpha \in b \exists \beta \in b\ (\alpha < \beta \wedge \alpha \text{ doesn't survive at } \beta).$$

Let $lh\ (\mathcal{B}_0) = \alpha_0 + 1$.

For each $\beta \leq \alpha_0$, letting $k = \deg^{\mathcal{B}}(\beta)$, choose a cofinal $Y_\beta \subseteq \rho_k(\mathcal{M}_\beta^{\mathcal{B}_0})$ such that $\pi_\beta^{\mathcal{E}_0}$ is $r\Sigma_{k+1}$ elementary on parameters from Y_β. Next, for $\alpha_0 < \beta < \theta$, we define Y_β by induction. If T-pred$(\beta) = \gamma$ and $\beta \notin D^T$ and $\deg(\beta) = \deg(\gamma)$, then set $Y_\beta = i_{\gamma\beta}{}''Y_\gamma$. If T-pred$(\beta) = \gamma$ but $\beta \in D^T$ or $\deg(\beta) < \deg(\gamma)$, then set $Y_\beta = i_\beta^*{}''(\mathcal{M}_\beta^*)$. Finally, if β is a limit ordinal $< \theta$, let $Y_\beta =$ common value of $i''_{\gamma\beta}Y_\gamma$ for all sufficiently large $\gamma T \beta$.

The idea here is that in a copying construction beginning from $\mathcal{E}_0$, Y_β is the subset of $\mathcal{M}_\beta$ on which we expect $r\Sigma_{k+1}$ elementarity of the copy map, for $k = \deg(\beta)$. We call a k realization $(\mathcal{R}, Q, \pi)$ of $\mathcal{M}_\beta$ a $(k, \boldsymbol{Y})$ *realization* just in case π is $r\Sigma_{k+1}$ elementary on Y_β. A realization Σ of $\Phi(\mathcal{T} \restriction \alpha+1)$ is a $\boldsymbol{Y}$ *realization* just in case $\forall \beta \leq \alpha((\mathcal{R}_\beta^{\mathcal{E}}, Q_\beta^{\mathcal{E}}, \pi_\beta^{\mathcal{E}})$ is a $(\deg(\beta), \boldsymbol{Y})$ realization of $\mathcal{M}_\beta)$. All realizations we consider in the proof to follow will be $\boldsymbol{Y}$-realizations.

Let $\alpha < \theta$ and let $(\mathcal{R}, Q, \pi)$ be a $\deg(\alpha)$ realization of $\mathcal{M}_\alpha$. We shall define a tree $U = U(\alpha, \mathcal{R}, Q, \pi)$. Roughly speaking, U tries to build a maximal branch b of T such that $\alpha \in b$, together with a realizing map σ for $\mathcal{M}_b^{\mathcal{T}}$ which extends π. More precisely, we put a triple
$(\langle \beta_0, \ldots, \beta_n \rangle, \langle \varphi_0, \ldots, \varphi_n \rangle, \langle Q_0, \ldots, Q_n \rangle)$ into U just in case

(1) $\beta_0 = \alpha$, $\varphi_0 = \pi$, and $Q_0 = Q$,

and for all $i < n$,

(2) $\beta_i T \beta_{i+1}$ and β_i does not survive at β_{i+1},

(3) $\text{ind}^{\mathcal{R}}(Q_{i+1}) \leq \text{ind}^{\mathcal{R}}(Q_i)$, and $D^{\mathcal{T}} \cap (\beta_i, \beta_{i+1}]_T \neq \phi$ iff $\text{ind}^{\mathcal{R}}(Q_{i+1}) < \text{ind}^{\mathcal{R}}(Q_i)$; moreover, if $D^{\mathcal{T}} \cap (\beta_i, \beta_{i+1}]_T = \phi$ and $\deg^{\mathcal{T}}(\beta_i) = \deg^{\mathcal{T}}(\beta_{i+1})$, then $Q_i = Q_{i+1}$, and

(4) $(\mathcal{R}, Q_{i+1}, \varphi_{i+1})$ is a $(\deg(\beta_{i+1}), \boldsymbol{Y})$ realization; moreover, if $D^{\mathcal{T}} \cap (\beta_i, \beta_{i+1}]_T \neq \phi$ and $\deg(\beta_i) = \deg(\beta_{i+1})$, then $\varphi_i = \varphi_{i+1} \circ i^{\mathcal{T}}_{\beta_i, \beta_{i+1}}$.

Suppose that $(\langle \beta_i \mid i \in \omega \rangle, \langle \varphi_i \mid i \in \omega \rangle, \langle Q_i \mid i \in \omega \rangle)$ is an infinite branch of $U(\alpha, \mathcal{R}, Q, \pi)$. Set $b = \{\eta \mid \exists i(\eta T \beta_i)\}$; then (1) and (2) guarantee that b is a maximal branch of $\mathcal{T}$ such that $\alpha \in b$. By condition (3), $D^{\mathcal{T}} \cap b$ is finite, and Q_i is eventually constant as $i \to \omega$, say with value Q_∞. Condition (3) also guarantees $Q_\infty = Q_0 = Q$ in the case $D^{\mathcal{T}} \cap (b - (\alpha+1)) = \phi$ and $\deg(\alpha) = \deg$ (b) (i.e., $\deg(\eta) = \deg\ (\alpha)$ for all $\eta \in b - (\alpha+1)$). Finally, let $y \in \mathcal{M}_b^{\mathcal{T}}$, and let k be large enough that $D^{\mathcal{T}} \cap (b - \beta_k) = \phi$, $\deg(\beta_k) = \deg$ (b), and $y = i^{\mathcal{T}}_{\beta_k b}(x)$ for some $x \in \mathcal{M}_{\beta_k}$. We then set $\sigma(y) = \varphi_k(x)$. By condition (4), σ is a well-defined weak deg (b) embedding from $\mathcal{M}_b$ into Q_∞. Moreover, if $D^{\mathcal{T}} \cap (b - (\alpha+1)) = \phi$ and $\deg\ (\alpha) = \deg(b)$, then $\pi = \sigma \circ i^{\mathcal{T}}_{\alpha b}$.

So if for some $\alpha \leq \alpha_0$, $U(\alpha, \mathcal{R}_\alpha^{\mathcal{E}_0}, Q_\alpha^{\mathcal{E}_0}, \pi_\alpha^{\mathcal{E}_0})$ has an infinite branch, then conclusion (1) of 9.14 holds. We therefore assume henceforth that for all $\alpha \leq \alpha_0$, $U(\alpha, \mathcal{R}_\alpha^{\mathcal{E}_0}, Q_\alpha^{\mathcal{E}_0}, \pi_\alpha^{\mathcal{E}_0})$ is wellfounded. Notice that $U(\alpha, \mathcal{R}_\alpha^{\mathcal{E}_0}, Q_\alpha^{\mathcal{E}_0}, \pi_\alpha^{\mathcal{E}_0})$ belongs to $\mathcal{R}_\alpha^{\mathcal{E}_0}$, and has size $< \delta^{\mathcal{R}_\alpha^{\mathcal{E}_0}}$ in $\mathcal{R}_\alpha^{\mathcal{E}_0}$.

Notice that if $\alpha < \theta$, then there are only finitely many $\gamma < \theta$ such that $\alpha \leq \gamma$ and T-pred$(\gamma) \leq \alpha$ and T-pred(γ) survives at γ. [If not, then we can fix $k < n(\alpha)$ such that $k = n(\gamma) = n(T\text{-pred}(\gamma))$ for infinitely many γ such that T-pred$(\gamma) < \alpha < \gamma$. Fix two distinct such γ's, say γ_0 and γ_1. Then γ_0 and γ_1 are T-incomparable, yet $n(\gamma_0) = n(\gamma_1)$. This contradicts the definition of n.] For $\alpha \leq \beta < \theta$, we define

$$c(\alpha, \beta) = |\{\gamma \mid \beta \leq \gamma < \theta \wedge \text{T-pred}(\gamma) \leq \alpha \ \wedge \text{T-pred}(\gamma) \text{ survives at } \gamma\}|.$$

Definition 9.15. *Let $\gamma < \theta = lh\ \mathcal{T}$, and let $\mathcal{E}$ be a realization of $\Phi(\mathcal{T} \restriction \gamma)$. We say $\mathcal{E}$ has enough room iff $\forall\, \alpha < \gamma$*

(a) *$\mathcal{U}(\alpha, \mathcal{R}^{\mathcal{E}}_{\alpha}, Q^{\mathcal{E}}_{\alpha}, \pi^{\mathcal{E}}_{\alpha})$ is wellfounded, and*

(b) *$\mathcal{R}^{\mathcal{E}}_{\alpha}$ has $\omega \cdot rank(U(\alpha, \mathcal{R}^{\mathcal{E}}_{\alpha}, Q^{\mathcal{E}}_{\alpha}, \pi^{\mathcal{E}}_{\alpha})) + c(\alpha, \gamma)$ cutoff points.*

Definition 9.16. *Let $\alpha < \gamma \leq \theta$; then α is a break point at γ iff whenever β is a successor ordinal such that $\alpha < \beta \leq \gamma$ and T-pred$(\beta) \leq \alpha$, then T-pred(β) does not survive at β.*

We can now prove our main lemma, which concerns the extendibility of realizations of the phalanxes determined by initial segments of $\mathcal{T}$.

Lemma 9.17. *Let $\alpha_0 \leq \alpha < \eta < \theta$, and let $\mathcal{E}$ be a realization of $\Phi(\mathcal{T} \restriction \alpha + 1)$ such that $\mathcal{E}$ has enough room. Then:*

(1) *Suppose α is a break point at η. Then there is a realization $\mathcal{F}$ of $\Phi(\mathcal{T} \restriction \eta + 1)$ such that $\mathcal{F} \restriction \alpha + 1 = \mathcal{E}$, $\mathcal{F}$ has enough room, and $\mathcal{R}^{\mathcal{F}}_{\eta} \in \mathcal{R}^{\mathcal{E}}_{\alpha}$.*

(2) *Suppose that for some $\delta \leq \alpha$, δ survives at η, and let δ_0 be the largest such ordinal δ. Then there is a realization $\mathcal{F}$ of $\Phi(\mathcal{T} \restriction \eta + 1)$ such that $\mathcal{F} \restriction \delta_0 = \mathcal{E} \restriction \delta_0$, $\mathcal{F}$ has enough room, and*

(a) *$\mathcal{R}^{\mathcal{F}}_{\eta} = \mathcal{R}^{\mathcal{E}}_{\delta_0}$ and $ind^{\mathcal{R}^{\mathcal{F}}_{\eta}}(Q^{\mathcal{F}}_{\eta}) \leq ind^{\mathcal{R}^{\mathcal{E}}_{\alpha}}(Q^{\mathcal{E}}_{\alpha})$,*

(b) *$D^{\mathcal{T}} \cap (\alpha, \eta]_T \neq \phi \Rightarrow ind^{\mathcal{R}^{\mathcal{F}}_{\eta}}(Q^{\mathcal{F}}_{\eta}) < ind^{\mathcal{R}^{\mathcal{E}}_{\alpha}}(Q^{\mathcal{E}}_{\alpha})$, and*

(c) *if $D^{\mathcal{T}} \cap (\alpha, \eta]_T = \phi$ and $deg^{\mathcal{T}}(\alpha) = deg^{\mathcal{T}}(\eta)$, then $Q^{\varepsilon}_{\alpha} = Q^{\mathcal{F}}_{\eta}$ and $\pi^{\varepsilon}_{\alpha} = \pi^{\mathcal{F}}_{\eta} \circ i^{\mathcal{T}}_{\alpha\eta}$.*

Proof. By induction on η. First, supposing 9.17 known for $\eta \leq \gamma$, we prove it for $\gamma + 1$. So let $\alpha_0 \leq \alpha < \gamma + 1$, and let $\mathcal{E}$ realize $\Phi(\mathcal{T} \restriction \alpha + 1)$ and have enough room. Let $\beta = T$-pred$(\gamma + 1)$.

We shall ultimately consider two cases in the construction of the desired $\mathcal{F}$ realizing $\Phi(\mathcal{T} \restriction \gamma + 2)$: the case that for some $\delta \leq \alpha$, δ survives at $\gamma + 1$, and the case that α is a break point at $\gamma + 1$ and β does not survive at $\gamma + 1$. Ostensibly there is a third case, the case that α is a break point at $\gamma + 1$ and β survives at $\gamma + 1$, but this case reduces easily to case one. For in this third case, $\alpha < \beta < \gamma + 1$. Since α is a break point at β, induction hypothesis 9.17 (1) gives us a $\mathcal{G}$ realizing $\Phi(\mathcal{T} \restriction \beta + 1)$, having enough room, and such that $\mathcal{E} = \mathcal{G} \restriction \alpha + 1$ and $\mathcal{R}^{\mathcal{G}}_{\beta} \in \mathcal{R}^{\mathcal{E}}_{\alpha}$. Now case one gives us an $\mathcal{F}$ realizing $\Phi(\mathcal{T} \restriction \gamma + 2)$, having enough room, and such that $\mathcal{F} \restriction \beta = \mathcal{G} \restriction \beta$ and $\mathcal{R}^{\mathcal{F}}_{\gamma+1} = \mathcal{R}^{\mathcal{G}}_{\beta}$. Clearly, $\mathcal{F}$ is as required in 9.17 (1) with $\eta = \gamma + 1$.

The desired $\mathcal{F}$ will come from a realization $\mathcal{H}$ of $\Phi(\mathcal{T} \restriction \gamma+1)$ which we now define. The definition depends on which of the two cases we are in.

Case 1. For some $\delta \leq \alpha$, δ survives at $\gamma+1$.

Let δ_0 be the largest such δ. Since $\beta = T\text{-pred}(\gamma+1)$, $\delta_0 = \beta$ or ($\delta_0 T \beta$ and δ_0 survives at β). Let $\mathcal{G} = \mathcal{E}$ if $\delta_0 = \beta$, and otherwise let $\mathcal{G}$ be the realization of $\Phi(\mathcal{T} \restriction \beta+1)$ given by our induction hypothesis 9.17 (2), with $\eta = \beta$. Since β survives at $\gamma+1$, either $\beta = \gamma$ or β is a break point at γ. [If $T\text{-pred}(\xi) \leq \beta < \xi \leq \gamma$ and $T\text{-pred}(\xi)$ survives at ξ, then $n(\beta) > n(\xi) = n(T\text{-pred}(\xi))$, so β doesn't survive at $\gamma+1$.] Let $\mathcal{H} = \mathcal{G}$ if $\beta = \gamma$, and otherwise let $\mathcal{H}$ be a realization of $\Phi(\mathcal{T} \restriction \gamma+1)$ such that $\mathcal{H} \restriction \beta+1 = \mathcal{G}$ as given by our induction hypothesis 9.17 (1), with $\eta = \gamma$.

Notice that in any case, $\mathcal{R}^{\mathcal{H}}_\beta = \mathcal{R}^{\mathcal{G}}_\beta = \mathcal{R}^{\mathcal{E}}_\alpha$. Also, if $D^{\mathcal{T}} \cap (\alpha,\beta]_T = \phi$ and $\deg^{\mathcal{T}}(\alpha) = \deg^{\mathcal{T}}(\beta)$, then $Q^{\mathcal{H}}_\beta = Q^{\mathcal{E}}_\alpha$ and $\pi^{\mathcal{E}}_\alpha = \pi^{\mathcal{H}}_\beta \circ i^{\mathcal{T}}_{\alpha\beta}$. Finally, $\text{ind}^{\mathcal{R}^{\mathcal{H}}_\beta}(Q^{\mathcal{H}}_\beta) \leq \text{ind}^{\mathcal{R}^{\mathcal{E}}_\alpha}(Q^{\mathcal{E}}_\alpha)$, and if $D^{\mathcal{T}} \cap (\alpha,\beta]_T \neq \phi$, then $\text{ind}^{\mathcal{R}^{\mathcal{H}}_\beta}(Q^{\mathcal{H}}_\beta) < \text{ind}^{\mathcal{R}^{\mathcal{H}}_\alpha}(Q^{\mathcal{H}}_\alpha)$.

Case 2. α is a break point at $\gamma+1$, and β does not survive at $\gamma+1$.

In this case, α is a break point at γ. If $\alpha = \gamma$ we let $\mathcal{H} = \mathcal{E}$. If $\alpha < \gamma$, we let $\mathcal{H}$ be the realization of $\Phi(\mathcal{T} \restriction \gamma+1)$ given by induction hypothesis 9.17 (1). In either case, we have $\mathcal{E} = \mathcal{H} \restriction \alpha+1$ and $\mathcal{R}^{\mathcal{H}}_\gamma \subseteq \mathcal{R}^{\mathcal{E}}_\alpha$.

Now, using $\mathcal{H}$, we produce the desired $\mathcal{F}$ realizing $\Phi(\mathcal{T} \restriction \gamma+2)$. We shall have to consider the case split above again later, but for now we can run the two cases simultaneously. In order to clean up our notation a bit, we set $(Q_\eta, \mathcal{R}_\eta, \pi_\eta) = (Q^{\mathcal{H}}_\eta, \mathcal{R}^{\mathcal{H}}_\eta, \pi^{\mathcal{H}}_\eta)$ for all $\eta \leq \gamma$.

Let $j = \deg(\gamma)$, and let Q_γ have index (j,ξ) in $\mathbb{C}^{\mathcal{R}_\gamma}$. Let $(\sigma^\gamma, \text{Res}^\gamma)$ be the complete resurrection of $\pi_\gamma(\lambda_\gamma)$ from (Q_γ, j, ξ), as computed in $\mathcal{R}_\gamma$, of course. Since $\gamma \geq \alpha_0$, $\lambda_\gamma = lh\ E^{\mathcal{T}}_\gamma$. If $\lambda_\gamma = \text{OR}^{\mathcal{M}_\gamma}$, then as usual we set $\pi_\gamma(\lambda_\gamma) = \text{OR}^{Q_\gamma}$.

Claim 1. If $\eta < \gamma$, then $\sigma^\gamma \restriction \pi_\gamma(\lambda_\eta) = \text{identity}$.

Proof. Since $\Phi(\mathcal{T} \restriction \gamma+1)$ is a phalanx, definition 9.6 guarantees that λ_η is a cardinal of $\mathcal{M}_\gamma$ and $\rho_j(\mathcal{M}_\gamma) > \lambda_\eta$. Since π_γ is a weak j-embedding, $\pi_\gamma(\lambda_\eta)$ is a cardinal of Q_γ and $\rho_j(Q_\gamma) > \pi_\gamma(\lambda_\eta)$. Also, $\pi_\gamma(\lambda_\eta) < \pi_\gamma(\lambda_\gamma)$. It follows that all projecta associated to the $(j, \pi_\gamma(\lambda_\gamma))$ dropdown sequence of Q_γ are $\geq \pi_\gamma(\lambda_\eta)$. □

Set

$$F = \sigma^\gamma \circ \pi_\gamma(E^{\mathcal{T}}_\gamma) = \text{ last extender of Res}^\gamma,$$

where if Res^γ is a bicephalus we choose the extender interpreting the same predicate symbol that E_γ interprets in $\mathcal{M}_\gamma$. We wish to consider $\text{Ult}(Q^*_{\gamma+1}, F)$, where $Q^*_{\gamma+1}$ is the creature of $\mathbb{C}^{\mathcal{R}_\beta}$ we shall now define. Let $n = \deg(\beta)$, and $\langle(\eta_0,k_0),\ldots,(\eta_e,k_e)\rangle$ = the (n,λ_β) dropdown sequence of $\mathcal{M}_\beta$, and set

$$\kappa_i = \rho_{k_i}(\mathcal{J}^{\mathcal{M}_\beta}_{\eta_i})$$

for $0 \leq i \leq e$. The following claim relates these to the $(n, \pi_\beta(\lambda_\beta))$ dropdown sequence of Q_β. The claim is slightly complicated by the fact that π_β is only a *weak* n-embedding.

Claim 2. The $(n, \pi_\beta(\lambda_\beta))$ dropdown sequence of Q_β is

(a) $\langle(\pi_\beta(\eta_0), k_0), \ldots, (\pi_\beta(\eta_e), k_e)\rangle$ if $\kappa_e < \rho_n(\mathcal{M}_\beta)$,

(b) $\langle(\pi_\beta(\eta_0), k_0), \ldots, (\pi_\beta(\eta_e), k_e)\rangle^\frown u$, where $u = \phi$ or $u = (\eta, n)$ for $\omega\eta = \mathrm{OR}^{Q_\beta}$ if $\kappa_e = \rho_n(\mathcal{M}_\beta)$ but $(\omega\eta_e, k_e) \neq (\mathrm{OR}^{\mathcal{M}_\beta}, n)$, and

(c) $\langle(\pi_\beta(\eta_0), k_0), \ldots, (\pi_\beta(\eta_{e-1}), k_{e-1})\rangle^\frown u$, where $u = \phi$ or $u = (\pi_\beta(\eta_e), k_e) = (\omega\eta, n)$, for $\omega\eta = \mathrm{OR}^{Q_\beta}$ if $(\omega\eta_e, k_e) = (\mathrm{OR}^{\mathcal{M}_\beta}, n)$.

Remark. Note that $\kappa_e = \rho_n(\mathcal{M}_\beta)$ in case (c). If $e = 0$, then $n = 0 = k_0$ and $\eta_0 = \lambda_\beta = \omega\lambda_\beta = \mathrm{OR}^{\mathcal{M}_\beta}$. The $(n, \pi_\beta(\lambda_\beta))$ dropdown sequence for Q_β is then $\langle(\mathrm{OR}^{Q_\beta}, 0)\rangle$, which falls under case (c).

Remark. The $u = \phi$ case in (c) would not be necessary if π_β were a full n-embedding.

The claim follows quite easily from the fact that π_β is a weak (n, Y_β)-embedding. For (a), notice that $\pi_\beta''\rho_n(\mathcal{M}_\beta) \leq \rho_n(Q_\beta)$. Recall that π_β preserves cardinals, so that if for example $\omega\eta_e < \mathrm{OR}^{\mathcal{M}_\beta}$ then $\mathcal{M}_\beta \models \forall\gamma \geq \eta_e(\rho_\omega(\mathcal{J}_\gamma^{\dot{E}}) \geq \rho_{k_e}(\mathcal{J}_{\eta_e}^{\dot{E}}))$, and thus $Q_\beta \models \forall\gamma \geq \pi_\beta(\eta_e)(\rho_\omega(\mathcal{J}_\gamma^{\dot{E}}) \geq \pi_\beta(\kappa_e))$.

Let $\mu_0 = \mathrm{crit}(E_\gamma^{\mathcal{T}})$, and let

$$i = \begin{cases} e+1 & \text{if} \quad \mu_0 < \kappa_e\,, \\ \text{least } j \text{ s.t. } \kappa_j \leq \mu_0\,, & \text{if } \kappa_e \leq \mu_0\,. \end{cases}$$

Notice that since $\kappa_0 = \lambda_\beta > \mu_0$, $i > 0$.

Because $\mathcal{T}$ is maximal,

$$\mathcal{M}_{\gamma+1}^* = \begin{cases} \mathcal{J}_{\eta_i}^{\mathcal{M}_\beta} & \text{if} \quad i \leq e\,, \\ \mathcal{M}_\beta & \text{if} \quad i = e+1\,, \end{cases}$$

and

$$\deg(\gamma+1) = \begin{cases} k_i - 1 & \text{if} \quad i \leq e\,, \\ n & \text{if} \quad i = e+1\,. \end{cases}$$

Let $(\sigma_i^\beta, \mathrm{Res}_i^\beta)$ be the ith partial resurrection of λ_β from (Q_β, n, τ), where Q_β has index (n, τ) in $\mathbb{C}^{\mathcal{R}_\beta}$, if this resurrection is defined. (The resurrection is undefined if $i = e+1$, and defined if $i < e$ by claim 2. If $i = e$, $(\sigma_i^\beta, \mathrm{Res}_i^\beta)$ is undefined just in case $(\omega\eta_e, k_e) = (\mathrm{OR}^{\mathcal{M}_\beta}, n)$ and the conclusion of (c) of claim 2 holds with $u = \phi$.)

Now let

$$Q^*_{\gamma+1} = \begin{cases} \mathrm{Res}^\beta_i & \text{if } \mathrm{Res}^\beta_i \text{ is defined},\\ Q_\beta & \text{otherwise} \end{cases}$$

$$\sigma = \begin{cases} \sigma^\beta_i & \text{if } \mathrm{Res}^\beta_i \text{ is defined},\\ \text{identity} & \text{otherwise}. \end{cases}$$

Thus, in any case, $\sigma \circ (\pi_\beta \restriction \mathcal{M}^*_{\gamma+1})$ is a weak $\deg(\gamma+1)$ embedding from $\mathcal{M}^*_{\gamma+1}$ into $Q^*_{\gamma+1}$. Moreover, $\sigma \circ \pi_\beta$ is $r\Sigma_{\deg(\gamma+1)+1}$ elementary on Z, where $Z = Y_\beta$ if $\gamma+1 \notin D^{\mathcal{T}}$ and $\deg(\gamma+1) = \deg(\beta)$, and $Z =$ universe of $\mathcal{M}^*_{\gamma+1}$ otherwise.

Set $k = \deg(\gamma+1)$.

Claim 2.5. $\sigma \circ \pi_\beta$ is a weak k-embedding which is $r\Sigma_{k+1}$ elementary on ${i^*_{\gamma+1}}^{-1}(Y_{\gamma+1})$.

Proof. Assume first that Res^β_i is defined, so that $i \leq e$, $\deg(\gamma+1) = k_i - 1$, and $\sigma = \sigma^\beta_i$ is a full $k_i - 1$ embedding. Looking at claim 2, we see that in all cases the domain of σ is $\mathcal{J}^{Q_\beta}_{\pi_\beta(\eta_i)}$, since we cannot have the situation in (c) with $i = e$ and $u = \phi$. But $\mathcal{M}^*_{\gamma+1} = \mathcal{J}^{\mathcal{M}_\beta}_{\eta_i}$, and $\pi_\beta \restriction \mathcal{M}^*_{\gamma+1}$ is a weak $k_i - 1$ embedding. In fact, if $\omega\eta_i < \mathrm{OR}^{\mathcal{M}_\beta}$, then $\pi_\beta \restriction \mathcal{M}^*_{\gamma+1}$ is fully elementary, and if $\omega\eta_i = \mathrm{OR}^{\mathcal{M}_\beta}$, then $k_i \leq n$, so $\pi_\beta \restriction \mathcal{M}^*_{\gamma+1}$ is a weak k_i embedding. It follows that $\sigma \circ (\pi_\beta \restriction \mathcal{M}^*_{\gamma+1})$ is a weak $k_i - 1$ embedding from $\mathcal{M}^*_{\gamma+1}$ into $Q^*_{\gamma+1}$. Assume next that Res^β_i is undefined. Then either $i = e+1$, or we have the situation in (c) of claim 2 with $u = \phi$. In either case, $\deg(\gamma+1) \leq n$. Also $\mathcal{M}^*_{\gamma+1} = \mathcal{M}_\beta$, $Q^*_{\gamma+1} = Q_\beta$, and σ=identity. Since π_β is a weak n-embedding, $\sigma \circ \pi_\beta$ is a weak $\deg(\gamma+1)$ embedding from $\mathcal{M}^*_{\gamma+1}$ into $Q^*_{\gamma+1}$.

Let $(\sigma^\beta, \mathrm{Res}^\beta)$ be the complete resurrection of $\pi_\beta(\lambda_\beta)$ from (Q_β, n, τ). Let ψ be the complete resurrection embedding for $\sigma(\pi_\beta(\lambda_\beta))$ from the appropriate tuple. (This tuple is $(Q^*_{\gamma+1}, n, \tau)$ if Res^β_i is undefined, and $(Q^*_{\gamma+1}, k_i - 1, \eta)$ where $\mathrm{Res}^\beta_i = \mathfrak{C}_{k_i-1}(\mathcal{N}_\eta)^{\mathcal{R}_\beta}$ otherwise.) Then

$$\psi : \mathcal{J}^{Q^*_{\gamma+1}}_{\sigma\circ\pi_\beta(\lambda_\beta)} \to \mathrm{Res}^\beta$$

and

$$\sigma^\beta = \psi \circ (\sigma \restriction \mathcal{J}^{Q_\beta}_{\pi_\beta(\lambda_\beta)}).$$

Claim 3. $\psi \restriction (\sup(\sigma \circ \pi_\beta'' \kappa_{i-1})) =$ identity.

Proof. Suppose first that Res^β_i exists, so that $i \leq e$ and $\sigma = \sigma^\beta_i$. From claim 2 and the fact that π_β is a weak n-embedding we see that $\pi_\beta(\kappa_{i-1})$ is the projectum associated to the $(i-1)^st$ element of the $(n, \pi_\beta(\lambda_\beta))$ drop-down sequence of Q_β. As we remarked earlier, ψ is therefore the identity on $\sup(\sigma^{\beta}_i{}'' \pi_\beta(\kappa_{i-1}))$, and this implies the claim.

Suppose next that Res^β_i is undefined, so that either $i = e+1$ or $i = e$ and (c) of claim 2 holds with $u = \phi$. In either case the projectum associated to the last term of the $(n, \pi_\beta(\lambda_\beta))$ dropdown sequence of Q_β is at least $\sup(\pi''_\beta \kappa_{i-1})$. Thus $\sigma^\beta \restriction \sup(\pi''_\beta \kappa_{i-1}) =$ identity. But $\psi = \sigma^\beta$ and $\sigma =$ identity, so this implies the claim. □

Now let

$$\mu_1 = \begin{cases} (\mu_0^+)^{\mathcal{M}^*_{\gamma+1}} & \text{if } \mathcal{M}^*_{\gamma+1} \models \mu_0^+ \text{ exists}, \\ \mathrm{OR}^{\mathcal{M}^*_{\gamma+1}} & \text{otherwise}. \end{cases}$$

Claim 4. $\mu_1 \leq \lambda_\beta$, and if $\mu_1 = \mathrm{OR}^{\mathcal{M}^*_{\gamma+1}}$ then $\mathcal{M}^*_{\gamma+1} = \mathcal{J}^{\mathcal{M}_\beta}_{\lambda_\beta}$ and μ_0 is the largest cardinal of $\mathcal{M}^*_{\gamma+1}$.

Proof. If $\beta = \gamma$, then $(\mu_0^+)^{\mathcal{J}^{\mathcal{M}_\gamma}_{\lambda_\gamma}}$ exists (is $< \lambda_\gamma$) since E_γ has index λ_γ on the $\mathcal{M}_\gamma$ sequence. Also, $\mathcal{M}^*_{\gamma+1}$ is the shortest initial segment of $\mathcal{M}_\gamma$ over which a subset of μ_0 not in $\mathcal{J}^{\mathcal{M}_\gamma}_{\lambda_\gamma}$ is definable. Thus $\mu_1 = (\mu_0^+)^{\mathcal{M}^*_{\gamma+1}} = (\mu_0^+)^{\mathcal{J}^{\mathcal{M}_\gamma}_{\lambda_\gamma}} < \lambda_\gamma$, and $\lambda_\gamma \leq \mathrm{OR}^{\mathcal{M}^*_{\gamma+1}}$, which yields the claim.

Now let $\beta < \gamma$. We have $\mu_0 < \nu_\beta \leq \lambda_\beta$, and λ_β is a cardinal of $\mathcal{M}_\gamma$. Also $P(\mu_0) \cap \mathcal{M}_\gamma = P(\mu_0) \cap \mathcal{J}^{\mathcal{M}_\gamma}_{\lambda_\beta} = P(\mu_0) \cap \mathcal{J}^{\mathcal{M}_\beta}_{\lambda_\beta} = P(\mu_0) \cap \mathcal{M}^*_{\gamma+1}$. It follows that $\mu_1 \leq \lambda_\beta$. If $\mu_1 = \mathrm{OR}^{\mathcal{M}^*_{\gamma+1}}$, then as $\lambda_\beta \leq \mathrm{OR}^{\mathcal{M}^*_{\gamma+1}}$, $\mathcal{M}^*_{\gamma+1} = \mathcal{J}^{\mathcal{M}_\beta}_{\lambda_\beta}$ and μ_0 is the largest cardinal of $\mathcal{M}^*_{\gamma+1}$. □

From the proof above we see that if $\beta < \gamma$, then $\mu_1 = (\mu_0^+)^{\mathcal{M}_\gamma}$. Also, claim 4 implies $\mu_1 \leq \kappa_{i-1}$. If $\kappa_{i-1} = \lambda_\beta$ this is obvious. Otherwise κ_{i-1} is a cardinal of $\mathcal{J}^{\mathcal{M}_\beta}_{\lambda_\beta}$, since it is a projectum of some $\mathcal{J}^{\mathcal{M}_\beta}_\eta$ with $\eta \geq \lambda_\beta$. Since $\mu_0 < \kappa_{i-1}$ by the choice of i, $\mu_1 \leq \kappa_{i-1}$.

The next claim shows that Res^γ and $Q^*_{\gamma+1}$ have the agreement required for an application of the shift lemma.

Claim 5. (a) Res^γ agrees with $Q^*_{\gamma+1}$ below $\sup(\sigma \circ \pi_\beta{}''\mu_1)$,

(b) $\sigma^\gamma \circ \pi_\gamma \restriction \mu_1 = \sigma \circ \pi_\beta \restriction \mu_1$.

Proof.

Subclaim A. $Q^*_{\gamma+1}$ and Res^β agree below $\sup(\sigma \circ \pi_\beta{}''\mu_1)$, and $\sigma \circ \pi_\beta \restriction \mu_1 = \psi \circ \sigma \circ \pi_\beta \restriction \mu_1$.

Proof. This follows at once from claim 3 and the fact that $\mu_1 \leq \kappa_{i-1}$.

Subclaim A yields claim 5 at once in the case $\beta = \gamma$, so let us assume $\beta < \gamma$.

Subclaim B. If $\beta < \gamma$, then Res^β and Q_γ agree below $\sup(\sigma \circ \pi_\beta{}''\mu_1)$, and $\psi \circ \sigma \circ \pi_\beta \restriction \mu_1 = \pi_\gamma \restriction \mu_1$.

Proof. Recall that $\psi \circ \sigma \circ \pi_\beta = \sigma^\beta \circ \pi_\beta$. This subclaim therefore follows at once from the fact that $\mathcal{H}$ is a realization of $\Phi(\mathcal{T} \restriction \gamma + 1)$; see clause 2 of 9.11. Notice here that $\mu_1 \leq \lambda_\beta$ by claim 4.

Subclaim C. If $\beta < \gamma$, then Q_γ and Res^γ agree below $\sup(\sigma \circ \pi_\beta{}''\mu_1)$, and $\pi_\gamma \restriction \mu_1 = \sigma^\gamma \circ \pi_\gamma \restriction \mu_1$.

Proof. $\mu_1 \leq \lambda_\beta$, and $\sigma \circ \pi_\beta \upharpoonright \mu_1 = \pi_\gamma \upharpoonright \mu_1$, so $\sup(\sigma \circ \pi_\beta{}''\mu_1) \leq \pi_\gamma(\lambda_\beta)$. By claim 1, Q_γ and Res^γ agree below $\pi_\gamma(\lambda_\beta)$, and σ^γ is the identity there.

Together, A, B, and C yield claim 5. □

Let us define

$$\kappa = \sigma^\gamma \circ \pi_\gamma(\mu_0) = \sigma \circ \pi_\beta(\mu_0) = \mathrm{crit}\ F\,.$$

Thus $(\kappa^+)^{Q^*_{\gamma+1}} = \sigma \circ \pi_\beta(\mu_1)$, with the usual understanding if $\mu_1 = \mathrm{OR}^{\mathcal{M}^*_{\gamma+1}}$.

Claim 6. Res^γ agrees with $Q^*_{\gamma+1}$ below $(\kappa^+)^{Q^*_{\gamma+1}} \leq (\kappa^+)^{\mathrm{Res}^\gamma}$.

Proof. We prove this slight strengthening of claim 5(a) in the same way that we proved 5(a). First, Res^β and $Q^*_{\gamma+1}$ agree below $(\kappa^+)^{Q^*_{\gamma+1}}$, and $(\kappa^+)^{Q^*_{\gamma+1}} \leq (\kappa^+)^{\mathrm{Res}^\beta}$. This is because $\mu_0 < \kappa_{i-1}$, so $\sigma \circ \pi_\beta(\mu_0) = \kappa < \mathrm{crit}\ \psi$, so $(\kappa^+)^{Q^*_{\gamma+1}} \leq \mathrm{crit}\ \psi$. This finishes the proof of claim 6 if $\beta = \gamma$, so suppose $\beta < \gamma$. Since $\mu_1 \leq \lambda_\beta$, and $(\kappa^+)^{\mathrm{Res}^\beta} = \sigma^\beta \circ \pi_\beta(\mu_1)$, and $\mathcal{H}$ is a realization, we have Res^β agrees with Q_γ below $(\kappa^+)^{\mathrm{Res}^\beta}$ and $(\kappa^+)^{\mathrm{Res}^\beta} \leq (\kappa^+)^{Q_\gamma}$. But Q_γ agrees with Res^γ below $\sigma^\gamma \circ \pi_\gamma(\lambda_\beta)$, and $(\kappa^+)^{Q_\gamma} \leq \sigma^\gamma \circ \pi_\gamma(\lambda_\beta)$, which completes the proof. □

Claim 7. $V_\kappa^{\mathcal{R}_\beta} = V_\kappa^{\mathcal{R}_\gamma}$.

Proof. $\mu_0 < \nu_\beta$ because $\mathcal{T}$ is an iteration tree, so $\kappa = \sigma \circ \pi_\beta(\mu_0) = \sigma^\beta \circ \pi_\beta(\mu_0) < \sigma^\beta \circ \pi_\beta(\nu_\beta)$. The claim now follows from the fact than $\mathcal{H}$ is a realization; cf. 9.11 (2) (a). □

Now Res^γ is a creature of $\mathbb{C}^{\mathcal{R}_\gamma}$ with an index of the form $(0,\eta)$ in $\mathbb{C}^{\mathcal{R}_\gamma}$. Therefore $\mathcal{R}_\gamma$ has background certificates for the countable fragments of F. Let

$$\begin{aligned}(N,G) = \quad & \text{some } (\sigma^\gamma \circ \pi_\gamma(\nu_\gamma),\ \mathrm{ran}(\sigma^\gamma \circ \pi_\gamma)) - \\ & \text{certificate for } F, \text{ as computed in } \mathcal{R}_\gamma\,.\end{aligned}$$

Since $\mathrm{Ult}(N,G)$ is closed under ω-sequences, $\sigma^\gamma \circ \pi_\gamma \upharpoonright \nu_\gamma \in \mathrm{Ult}(N,G)$. Let us fix $b \in [lh\ G]^{<\omega}$ and a function $\bar{u} \mapsto \pi(\bar{u})$ mapping $[\kappa]^{|b|}$ into $V_\kappa^{\mathcal{R}_\gamma}$ so that

$$\sigma^\gamma \circ \pi_\gamma \upharpoonright (\lambda_\gamma + 1) = [b, \lambda\bar{u} \cdot \pi(\bar{u})]^N_G\,.$$

Suppose for a moment that case 1 of 9.17 applies, that is, that $\delta_0 \leq \alpha$ and δ_0 survives at $\gamma+1$. It follows that $c(\eta,\gamma+2) < c(\eta,\gamma+1)$ for all η such that $\beta \leq \eta \leq \gamma$. Therefore, for such η, $\mathcal{R}_\eta$ has $\omega \cdot \mathrm{rank}(\mathcal{U}(\eta,\mathcal{R}_\eta,Q_\eta,\pi_\eta)) + c(\eta,\gamma+2) + 1$ cutoff points, because $\mathcal{H}$ has enough room. Let ξ_η be the last of these cutoff points, and set

$$\begin{aligned}\mathcal{R}^*_\eta \ = \ & \text{transitive collapse of} \\ & \mathrm{Hull}^{V_{\xi_\eta}}(V_{\sigma^\eta \circ \pi_\eta(\nu_\eta)} \cup \{\delta^{\mathcal{R}_\eta}, Q_\eta, \pi_\eta\} \cup \sigma^\eta \circ \pi_\eta(\lambda_\eta))\,, \\ & \text{as computed in } \mathcal{R}_\eta\,,\end{aligned}$$

and

$$(Q^*_\eta, \pi^*_\eta) = \text{image of } (Q_\eta, \pi_\eta) \text{ under collapse}.$$

Notice that $\mathcal{R}^*_\eta$ is coded by an element of $V^{\mathcal{R}_\eta}_{\sigma^\eta \circ \pi_\eta(\nu_\eta)+1}$, which is a subset of $\mathcal{R}_\gamma$ because $\mathcal{H}$ is a realization. (Note here $\sigma^\eta \circ \pi_\eta(\lambda_\eta)$ has cardinality $\sigma_\eta \circ \pi_\eta(\nu_\eta)$ in $\mathcal{R}_\eta$.) So $(\mathcal{R}^*_\eta, Q^*_\eta, \pi^*_\eta) \in \mathcal{R}_\gamma$ for all η such that $\beta \leq \eta \leq \gamma$. Set

$$\mathcal{H}^* = \langle (\mathcal{R}^*_\eta, Q^*_\eta, \pi^*_\eta) \mid \beta \leq \eta \leq \gamma \rangle .$$

Then $\mathcal{H}^* \in \mathcal{R}_\gamma$ since $\mathcal{R}_\gamma$ is closed under ω sequences. Clearly, $\mathcal{H}^*$ is coded by a member of $V^{\mathcal{R}_\gamma}_{\sigma^\gamma \circ \pi_\gamma(\nu_\gamma)+1}$. It is easy to check that $(\mathcal{H} \upharpoonright \beta)^\frown \mathcal{H}^*$ is a realization of $\Phi(\mathcal{T} \upharpoonright \gamma+1)$. It may not have enough room as a realization of $\Phi(\mathcal{T} \upharpoonright \gamma+1)$, of course, because we have dropped an ordinal on coordinates η such that $\beta \leq \eta \leq \gamma$. Since G is $\sigma^\gamma \circ \pi_\gamma(\nu_\gamma) + 1$ strong in $\mathcal{R}_\gamma$, $\mathcal{H}^* \in \mathrm{Ult}(N, G)$. We may suppose our finite support b was chosen so that for some function $\bar{u} \mapsto \mathcal{H}^*(\bar{u})$ mapping $[\kappa]^{|b|}$ into V_κ,

$$\mathcal{H}^* = [b, \lambda \bar{u} \cdot \mathcal{H}^*(\bar{u})]^N_G .$$

If there is no $\delta \leq \alpha$ such that δ survives at $\gamma + 1$, then $\mathcal{H}^*$ is undefined.

Let $k = \deg(\gamma + 1)$, and $Q'_{\gamma+1} = \mathrm{Ult}_k(Q^*_{\gamma+1}, F)$. The ultrapower makes sense by claim 6, and it is wellfounded because F has background certificates in $\mathcal{R}_\gamma$, and $\mathcal{R}_\gamma$ is ω-closed. Let $\tau : \mathcal{M}_{\gamma+1} \to Q'_{\gamma+1}$ be given by the shift lemma, that is,

$$\tau\left([a, f]^{\mathcal{M}^*_{\gamma+1}}_{E_\gamma}\right) = [\sigma^\gamma \circ \pi_\gamma(a), \sigma \circ \pi_\beta(f)]^{Q^*_{\gamma+1}}_F .$$

(Here, if $k > 0$, then $\sigma \circ \pi_\beta(f_{r,q}) = f_{r,\sigma\circ\pi_\beta(q)}$ for all terms $r \in Sk_k$ and $q \in \mathcal{M}^*_{\gamma+1}$. For simplicity, we shall use the $k = 0$ ultrapower notation.) By the shift lemma, $Q'_{\gamma+1}$ agrees with Q_γ below $\sigma^\gamma \circ \pi_\gamma(\lambda_\gamma)$, and $\tau \upharpoonright \lambda_\gamma = \sigma^\gamma \circ \pi_\gamma \upharpoonright \lambda_\gamma$. Also, τ is a weak k-embedding which is $r\,\Sigma_{\sigma+1}$ elementary on $Y_{\gamma+1}$. We now use the countable completeness of G to reflect τ below κ.

Let $\{x_n \mid n < \omega\}$ be an enumeration of the universe of $\mathcal{M}_{\gamma+1}$, and let $x_n = [\bar{a}_n, \bar{f}_n]^{\mathcal{M}^*_{\gamma+1}}_{E_\gamma}$ where $\bar{a}_n \in [\nu_\gamma]^{<\omega}$. Set $a_n = \sigma^\gamma \circ \pi_\gamma(\bar{a}_n)$ and $f_n = \sigma \circ \pi_\beta(\bar{f}_n)$, so that

$$\tau(x_n) = [a_n, f_n]^{Q^*_{\gamma+1}}_F .$$

For notational reasons, we shall sometimes regard the component measures E_c of an extender E as concentrating on order-preserving $t : c \to \mathrm{crit}(E)$, so that "for E a.e. $t : c \to \mathrm{crit}(E)$, $t \in X$" means that there is a set $Y \in E_c$ such that whenever $t : c \to \mathrm{crit}(E)$ is order preserving and $t''c \in Y$, then $t \in X$. Let us write $I(\beta, Q, \sigma)$ just in case σ is $r\Sigma_k$ elementary on its domain, σ is $r\,\Sigma_{k+1}$ elementary on dom $\sigma \cap Y_\beta$, $\forall i < k(\rho_i(\mathcal{M}_\beta) \in \mathrm{dom}\,\sigma \Rightarrow \sigma(\rho_i(\mathcal{M}_\beta)) = \rho_i(Q))$, and $\sigma''\rho_k(\mathcal{M}_\beta) \subseteq \rho_k(Q)$. Thus, if $\pi : \mathcal{M}_\beta \to Q$, then π is a weak k -embedding from $\mathcal{M}_\beta$ into Q which is $r\,\Sigma_{k+1}$ elementary on $Y_\beta \Leftrightarrow (\forall \text{ finite } F \subseteq \mathcal{M}_\beta) I(\beta, Q, \pi \upharpoonright F)$.

For $t : (b \cup a_0 \cdots \cup a_n) \to \kappa$ order preserving, let

$$\varphi_t^n(x_i) = f_i(t''a_i)$$

for all $i \le n$.

Claim 8. Let $n < \omega$ and let $c = b \cup a_0 \cup \cdots \cup a_n$. Then there is a set $W_n \in G_c$ such that whenever $t : c \to \kappa$ is order preserving and $t''c \in W_n$,

(i) $I(\gamma + 1, Q^*_{\gamma+1}, \varphi_t^n)$
(ii) if $i^*_{\gamma+1}(y) \in \operatorname{dom} \varphi_t^n$, then $\varphi_t^n(i^*_{\gamma+1}(y)) = \sigma \circ \pi_\beta(y)$,
(iii) if $x_n < \lambda_\gamma$, then $\varphi_t^n(x_n) = \pi(t''b)(x_n)$, and
(iv) if $x_n = \lambda_\gamma$, then $\varphi_t^n(x_n) \ge \pi(t''b)(x_n)$.

Proof. We first show that (ii) holds for G a.e. $t : c \to \kappa$. Let $i^*_{\gamma+1}(y) = x_i = [\bar{a}_i, \bar{f}_i]_{E_\gamma}^{\mathcal{M}^*_{\gamma+1}}$, where $i \le n$. Then $\bar{f}_i(\bar{u}) = y$ for $(E_\gamma)_{\bar{a}_i}$ a.e. $\bar{u}$. Since $\sigma \circ \pi_\beta$ and $\sigma^\gamma \circ \pi_\gamma$ agree on $P(\mu_0)$, this means that $f_i(\bar{u}) = \sigma \circ \pi_\beta(y)$ for F_{a_i} a.e. $\bar{u}$. The set of such $\bar{u}$ is in $\operatorname{ran}(\sigma^\gamma \circ \pi_\gamma)$, so $f_i(t''a_i) = \sigma \circ \pi_\beta(y)$ for G a.e. $t : c \to \kappa$. Since $\varphi_t^n(i^*_{\gamma+1}(y)) = f_i(t''a_i)$, we are done.

Next, we show (i) holds G a.e. First, let $\rho(v_0 \cdots v_n)$ be an $r\Sigma_k$ formula. Then

$$\begin{aligned} \mathcal{M}_{\gamma+1} \models \rho[x_0 \cdots x_n] \quad & \text{iff } Q'_{\gamma+1} \models \rho[\tau(x_0) \cdots \tau(x_n)] \\ & \text{iff for } F \text{ a.e. } t : \bigcup_{i \le n} a_i \to \kappa, \\ & \quad Q^*_{\gamma+1} \models \rho[f_0(t''a_0) \cdots f_n(t''a_n)] \\ & \text{iff for } G \text{ a.e. } t : c \to \kappa, \\ & \quad Q^*_{\gamma+1} \models \rho[\varphi_t^n(x_0) \cdots \varphi_t^n(x_n)]\,. \end{aligned}$$

Notice, for the third equivalence above, that the appropriate set of $\bar{u}$ is in the range of $\sigma^\gamma \circ \pi_\gamma$, so that $F_{\bigcup_i a_i}$ and $G_{\bigcup_i a_i}$ give it the same measure. Second, we show φ_t^n is $r\Sigma_{k+1}$ elementary on $Y_{\gamma+1} \cap \{x_0 \cdots x_n\}$, for G a.e. $t : c \to \kappa$. Notice here that $Y_{\gamma+1} = i^{*\prime\prime}_{\gamma+1} Z$, where $\sigma \circ \pi_\beta$ is $r\Sigma_{k+1}$ elementary on Z. [If $\gamma + 1 \notin D^\mathcal{T}$ and $k = \deg(\gamma + 1) = \deg(\beta)$, then $Z = Y_\beta$, and $\sigma \circ \pi_\beta = \pi_\beta$ is $r\Sigma_{k+1}$ elementary on Y_β because $\mathcal{H}$ is a $\boldsymbol{Y}$-realization. Otherwise, Z is the universe of $\mathcal{M}^*_{\gamma+1}$, σ is a full k-embedding, and π_β is at least $r\Sigma_{k+1}$ as a map from $\mathcal{M}^*_{\gamma+1}$ to $\mathcal{J}_\eta^{Q_\beta}$, where $\omega\eta = \pi_\beta(\mathrm{OR} \cap \mathcal{M}^*_{\gamma+1})$.] Thus, if we set $Y_{\gamma+1} \cap \{x_0 \cdots x_n\} = \{i^*_{\gamma+1}(y_0), \cdots, i^*_{\gamma+1}(y_m)\}$, then we have for all $r\Sigma_{k+1}$ formulae ρ

$$\begin{aligned} & \mathcal{M}_{\gamma+1} \models \rho[i^*_{\gamma+1}(y_0) \cdots i^*_{\gamma+1}(y_m)] \text{ iff } \mathcal{M}^*_{\gamma+1} \models \rho[y_0 \cdots y_m] \\ & \quad \text{iff } Q^*_{\gamma+1} \models \rho[\sigma \circ \pi_\beta(y_0) \cdots \sigma \circ \pi_\beta(y_m)] \\ & \quad \text{iff for } G \text{ a.e. } t : c \to \kappa \\ & \qquad Q^*_{\gamma+1} \models \rho[\varphi_t^n(i^*_{\gamma+1}(y_0)) \cdots \varphi_t^n(i^*_{\gamma+1}(y_m))]\,. \end{aligned}$$

This completes the proof of (i).

We now prove (iii). Let $x_n < \lambda_\gamma$, and assume first that $\lambda_\gamma = \nu_\gamma$. Since $x_n < \nu_\gamma$,

$$[\bar{a}_n, \bar{f}_n]_{E_\gamma}^{\mathcal{M}^*_{\gamma+1}} = [\{x_n\}, \mathrm{id}]_{E_\gamma}^{\mathcal{M}^*_{\gamma+1}},$$

so

$$[a_n, f_n]_F^{Q^*_{\gamma+1}} = [\{\sigma^\gamma \circ \pi_\gamma(x_n)\}, \mathrm{id}]_F^{Q^*_{\gamma+1}},$$

because of the agreement between $\sigma \circ \pi_\beta$ and $\sigma^\gamma \circ \pi_\gamma$. Letting $d = a_n \cup \{\sigma^\gamma \circ \pi_\gamma(x_n)\}$, this means that for F a.e. $t : d \to \kappa$, $f_n(t''a_n) = t(\sigma^\gamma \circ \pi_\gamma(x_n))$. Because the set of all $t''d$ for which this equation holds is in $\mathrm{ran}(\sigma^\gamma \circ \pi_\gamma)$, we get that

$$f_n(t''a_n) = t(\sigma^\gamma \circ \pi_\gamma(x_n)), \text{ for } G \text{ a.e. } t.$$

But also,

$$[\{\sigma^\gamma \circ \pi_\gamma(x_n)\}, \mathrm{id}]_G^N = \sigma_\gamma \circ \pi_\gamma(x_n) = [b, \lambda\bar{u} \cdot \pi(\bar{u})]_G^N\ (x_n),$$

and since x_n is countable in N, it is represented by the constantly x_n function in $\mathrm{Ult}(N, G)$. By Los' theorem for $\mathrm{Ult}(N, G)$,

$$\pi(t''b)(x_n) = t(\sigma^\gamma \circ \pi_\gamma(x_n)), \text{ for } G \text{ a.e. } t.$$

This finishes the proof of (iii) in case $\lambda_\gamma = \nu_\gamma$.

If $\nu_\gamma < \lambda_\gamma$, then $\nu_\gamma = \nu + 1$ where ν is the largest generator of E_γ, and $\lambda_\gamma = lh\ E_\gamma = (\nu^+)^{\mathrm{Ult}(\mathcal{M}^*_{\gamma+1}, E_\gamma)}$.

If $x_n < \nu_\gamma$, the proof in the first case applies, so assume $x_n \geq \nu_\gamma$. We then get a function $\bar{g} \in V_{\mu_0+1}^{\mathcal{M}^*_{\gamma+1}}$ such that

$$[\bar{a}_n, \bar{g}]_{E_\gamma}^{\mathcal{M}^*_{\gamma+1}} = \text{some wellorder of } \nu \text{ of order type } x_n.$$

Applying the shift lemma map τ to this fact, with $g = \sigma \circ \pi_\beta(\bar{g}) = \sigma^\gamma \circ \pi_\gamma(\bar{g})$,

$$[a_n, g]_F^{Q^*_{\gamma+1}} = \text{some wellorder of } \sigma^\gamma \circ \pi_\gamma(\nu) \text{ of order type } \sigma^\gamma \circ \pi_\gamma(x_n).$$

But now F_d agrees with G_d on all sets in $\mathrm{ran}(\sigma^\gamma \circ \pi_\gamma)$, whenever $d \in [\sigma^\gamma \circ \pi_\gamma(\nu + 1)]^{<\omega}$. This implies

$$[a_n, g]_F^{Q^*_{\gamma+1}} = [a_n, g]_G^N.$$

It follows that for G a.e. t, $g(t''a_n)$ is a wellorder of order type $t(\sigma^\gamma \circ \pi_\gamma(x_n))$. We also have that for F a.e. t, hence for G a.e. t, $g(t''a_n)$ has order type $f(t''a_n)$. So we get that $f(t''a_n) = t(\sigma^\gamma \circ \pi_\gamma(x_n))$ for G a.e. t. Now we can finish the proof of (iii) as in the first case.

We leave the proof of (iv) to the reader. The main point is that $[a_n, f_n]_G^N \geq \sigma^\gamma \circ \pi_\gamma(\lambda_\gamma)$.

(We may assume $\bar{f}_n \in V_{\mu_0+1}^{\mathcal{M}^*_{\gamma+1}}$, so that $f_n = \sigma^\gamma \circ \pi_\gamma(\bar{f}_n)$ is in $\mathrm{ran}(\sigma^\gamma \circ \pi_\gamma)$.) This follows from the agreement between F and G; the proof breaks into the cases $\nu_\gamma = \lambda_\gamma$ and $\nu_\gamma < \lambda_\gamma$ as did the proof of (iii).

This completes the proof of claim 8.

We can now finish the proof of 9.17 in case 1, the case that for some $\delta \leq \alpha$, α survives at $\gamma+1$. For $\beta \leq \eta \leq \gamma$, let σ^*_η be the complete resurrection embedding in $\mathcal{R}^*_\eta$ for Q^*_η from $\pi^*_\eta(\lambda_\eta)$. Then $\sigma^*_\eta \restriction \lambda_\eta = \sigma^\gamma \circ \pi_\gamma \restriction \lambda_\eta$, and $\sigma^*_\eta \circ \pi^*_\eta(\lambda_\eta) \leq \sigma^\gamma \circ \pi_\gamma(\lambda_\eta)$ for all $\eta \leq \gamma$; this one sees from the construction of $\mathcal{H}^*$. This agreement is a fact about $\mathcal{H}^*$ and $\sigma^\gamma \circ \pi_\gamma \restriction (\lambda_\gamma + 1)$ in $\mathrm{Ult}(N, G)$; by Los' theorem we get a set $X \in G_b$ such that for all $\bar{u} \in X$

$$(\sigma^*_\eta \circ \pi^*_\eta)(\bar{u}) \restriction \lambda_\eta = \pi(\bar{u}) \restriction \lambda_\eta$$

and

$$(\sigma^*_\eta \circ \pi^*_\eta)(\bar{u})(\lambda_\eta) \leq \pi(\bar{u})(\lambda_\eta)$$

for $\beta \leq \eta \leq \gamma$. Here $(\sigma^*_\eta \circ \pi^*_\eta)(\bar{u}) = \sigma^*_\eta(\bar{u}) \circ \pi^*_\eta(\bar{u})$, where $\mathcal{H}^*_\eta(\bar{u}) = (\mathcal{R}^*_\eta(\bar{u}), Q^*_\eta(\bar{u}), \pi^*_\eta(\bar{u}))$ and $\sigma^*_\eta(\bar{u})$ is the complete resurrection of $\pi^*_\eta(\bar{u})(\lambda_\eta)$ from $Q^*_\eta(\bar{u})$ in $\mathcal{R}^*_\eta(\bar{u})$. We can also arrange that for $\bar{u} \in X$,

$$\sigma^\gamma \circ \pi_\gamma \restriction \mu_0 = \pi(\bar{u}) \restriction \mu_0\,,$$

because $\sigma^\gamma \circ \pi_\gamma{}'' \mu_0$ is just a countable subset of $\kappa = \mathrm{crit}\ G$, and G is countably complete. Finally, we can arrange that for $\bar{u} \in X$, $\mathcal{R}^*_\eta(\bar{u})$ has $\omega \cdot \mathrm{rank}(\mathcal{U}(\eta, \mathcal{R}^*_\eta(\bar{u}), Q^*_\eta(\bar{u}), \pi^*_\eta(\bar{u})) + c(\eta, \gamma+2)$ cutoff points.

Now let W_n be as in claim 8, for all $n < \omega$, and let $t : b \cup \bigcup_{n<\omega} a_n \to \kappa$ be order preserving and such that $t''b \in X$ and $t''(b \cup a_0 \cdots \cup a_n) \in W_n$ for all n. Such a t exists because G is countably complete. Set $\varphi(x_n) = f_n(t''a_n)$ for all $n < \omega$, and

$$\mathcal{F} = \mathcal{H} \restriction \beta ^\frown \mathcal{H}^*(t''b) ^\frown \langle (\mathcal{R}_\beta, Q^*_{\gamma+1}, \varphi) \rangle\,.$$

It is easy to verify that $\mathcal{F}$ fulfills the requirements of 9.17 as a realizations of $\Phi(\mathcal{T} \restriction \gamma+2)$ in case 1.

Now let us prove 9.17 in case 2, the case that α is a break point at $\gamma+1$ and β does not survive at $\gamma+1$. From claim 8 and the countable completeness of G we get

Claim 9. For G_b a.e. $\bar{u}$, there is a ($\deg(\gamma+1), Y)$ embedding $\varphi : \mathcal{M}_{\gamma+1} \to Q^*_{\gamma+1}$ such that

(a) $\varphi \restriction \lambda_\gamma = \pi(\bar{u}) \restriction \lambda_\gamma$,
(b) $\varphi(\lambda_\gamma) \geq \pi(\bar{u})(\lambda_\gamma)$,
(c) $\varphi \circ i^*_{\gamma+1} = \sigma \circ \pi_\beta$.

Now if $\bar{u}$ and φ are as in claim 9, then $(\gamma+1, \varphi, Q^*_{\gamma+1})$ is a node of the tree $\mathcal{U}(\beta, \mathcal{R}^{\mathcal{H}}_\beta, Q^{\mathcal{H}}_\beta, \pi^{\mathcal{H}}_\beta)$. (The fact that β does not survive at $\gamma+1$ is relevant here.) Moreover, $\mathcal{U}(\gamma+1, \mathcal{R}^{\mathcal{H}}_\beta, Q^*_{\gamma+1}, \varphi)$ is isomorphic to the subtree of $\mathcal{U}(\beta, \mathcal{R}_\beta^\mathcal{H}, Q_\beta^\mathcal{H}, \pi_\beta^\mathcal{H})$ consisting of nodes below $(\gamma+1, \varphi, Q_{\gamma+1}^*)$. It follows that $\mathcal{R}_\beta^\mathcal{H}$ has an $\omega \cdot \text{rank}(\mathcal{U}(\gamma+1, \mathcal{R}_\beta^\mathcal{H}, Q_{\gamma+1}^*, \varphi)) + c(\gamma+1, \gamma+2) + 1^{\text{st}}$ cutoff point η. Working in $\mathcal{R}_\beta^\mathcal{H}$, we can form a Skolem hull of $V_\eta^{\mathcal{R}_\beta^\mathcal{H}}$ containing $V_{\pi(\bar{u})(\lambda_\gamma)}^{\mathcal{R}_\beta^\mathcal{H}} \cup \{Q_{\gamma+1}^*\varphi\}$, closed under ω-sequences and having size $< \kappa$. The collapse of this hull belongs to $V_\kappa^{\mathcal{R}_\beta^\mathcal{H}} = V_\kappa^N$. This gives us

Claim 10. For G_b a.e. $\bar{u}$, there is a triple $(\mathcal{R}, Q, \varphi)$ such that

(a) $(\mathcal{R}, Q, \varphi)$ is a $(\deg(\gamma+1), Y)$ realization of $\mathcal{M}_{\gamma+1}$,

(b) $(\mathcal{R}, Q, \varphi) \in V_\kappa^N$,

(c) $\mathcal{R}$ has $\omega \cdot \text{rank}(\mathcal{U}(\gamma+1, \mathcal{R}, Q, \varphi)) + c(\gamma+1, \gamma+2)$ cutoff points,

(d) Q agrees with Q_γ below $\pi(\bar{u})(\lambda_\gamma)$, and $\mathcal{R}$ agrees with N below $\pi(\bar{u})(\lambda_\gamma)$,

and

(e) $\varphi \restriction \lambda_\gamma = \pi(\bar{u}) \restriction \lambda_\gamma$ and $\varphi(\lambda_\gamma) \geq \pi(\bar{u})(\lambda_\gamma)$.

By the axiom of choice in N, there is in N a function $f(\bar{u}) = (\mathcal{R}(\bar{u}), Q(\bar{u}), \varphi(\bar{u}))$ which picks, for each $\bar{u}$ in the relevant G_b-measure one set, a triple satisfying claim 10. Let

$$\mathcal{F} = \mathcal{H}^\frown[b, \lambda\bar{u} \cdot (\mathcal{R}(\bar{u}), Q(\bar{u}), \varphi(\bar{u}))]_G^N.$$

It is easy to see that $\mathcal{F}$ is a realization of the phalanx $\Phi(\mathcal{T} \restriction \gamma+2)$; the necessary agreement of models and embeddings comes from parts (d) and (e) of claim 10. Part (c) of claim 10 implies that $\mathcal{F}$ has enough room. As case 2 governed our definition of $\mathcal{H}$, $\mathcal{E} \restriction \alpha+1 = \mathcal{H} \restriction \alpha+1$ and $\mathcal{R}_\gamma^\mathcal{H} \subseteq \mathcal{R}_\alpha^\mathcal{E}$. It follows that $\mathcal{F} \restriction \alpha+1 = \mathcal{E} \restriction \alpha+1$, and since $\mathcal{R}_{\gamma+1}^\mathcal{F} \in \text{Ult}(N, G) \subseteq \mathcal{R}_\gamma^\mathcal{H}$, we have $\mathcal{R}_{\gamma+1}^\mathcal{F} \in \mathcal{R}_\alpha^\mathcal{E}$. Thus $\mathcal{F}$ witnesses the truth of 9.17 (1).

This finishes the successor step in the inductive proof of 9.17.

Now let η be a limit ordinal, and $\alpha_0 \leq \alpha < \eta$. Let $\beta T \eta$, where β is large enough that $\alpha < \beta$, β survives at η, $D^\mathcal{T} \cap [\beta, \eta]_T = \phi$, and $\deg(\beta) = \deg(\eta)$. Let $\langle \beta_n \mid n \in \omega \rangle$ be such that $\beta_0 = \beta$, and $\beta_n T \beta_{n+1} T \eta$ for all n, and $\eta = \sup\{\beta_n \mid n \in \omega\}$. Let $\mathcal{E}$ be our given realization of $\Phi(\mathcal{T} \restriction \alpha+1)$.

Suppose first that α is a break point at η. Then α is a break point at β, so by induction we have a realization $\mathcal{F}_0$ of $\Phi(\mathcal{T} \restriction \beta+1)$ which has enough room. We also get $\mathcal{F}_0 \restriction \alpha+1 = \mathcal{E}$, and $\mathcal{R}_\beta^{\mathcal{F}_0} \in \mathcal{R}_\alpha^\mathcal{E}$. Now suppose $\mathcal{F}_n$ realizing $\Phi(\mathcal{T}) \restriction (\beta_n+1)$ is given. Since β_n survives at β_{n+1}, and between β_n and β_{n+1} there is no dropping in model or degree, our induction hypothesis gives a realization $\mathcal{F}_{n+1}$ of $\Phi(\mathcal{T} \restriction \beta_{n+1}+1)$ having enough room, and such that $\mathcal{F}_{n+1} \restriction \beta_n = \mathcal{F}_n \restriction \beta_n$, $\mathcal{R}_{\beta_n}^{\mathcal{F}_n} = \mathcal{R}_{\beta_{n+1}}^{\mathcal{F}_{n+1}}$ and $Q_{\beta_n}^{\mathcal{F}_n} = Q_{\beta_{n+1}}^{\mathcal{F}_{n+1}}$, and $\pi_{\beta_n}^{\mathcal{F}_n} = \pi_{\beta_{n+1}}^{\mathcal{F}_{n+1}} \circ i_{\beta_n \beta_{n+1}}^\mathcal{T}$. Let

$$\mathcal{F} = \bigcup_n \mathcal{F}_n \restriction \beta_n{}^\frown \langle \mathcal{R}_\beta^{\mathcal{F}_0} Q_\beta^{\mathcal{F}_0}, \pi \rangle,$$

where for $x \in \mathcal{M}_\eta$ we define $\pi(x)$ by

$$x = i^{\mathcal{T}}_{\beta_n,\eta}(y) \Rightarrow \pi(x) = \pi^{\mathcal{F}_n}_{\beta_n}(y).$$

It is easy to check that $\mathcal{F}$ witnesses 9.17 (1) for α and η.

Next, suppose $\delta_0 \leq \alpha$ is largest such that δ_0 survives at η. Let $\langle \beta_n \mid n < \omega \rangle$ be such that T-pred$(\beta_0) = \delta_0$, $\beta_n T \beta_{n+1} T \eta$ for all n, and $\sup_n \beta_n = \eta$. By induction hypothesis 9.17 (2) we get a realization $\mathcal{F}_0$ of $\Phi(\mathcal{T} \restriction \beta_0 + 1)$ which has enough room, and such that $\mathcal{F}_0 \restriction \delta_0 = \mathcal{E} \restriction \delta_0$ and $\mathcal{R}^{\mathcal{F}_0}_{\beta_0} = \mathcal{R}^{\mathcal{E}}_{\delta_0}$, and $Q^{\mathcal{F}_0}_{\beta_0}$ is related as required to $Q^{\mathcal{E}}_{\delta_0}$, and $\pi^{\mathcal{E}}_{\delta_0} = \pi^{\mathcal{F}_0}_{\beta_0} \circ i^{\mathcal{T}}_{\delta_0,\beta_0}$ if this is required. We can then use induction hypothesis 9.17 (2) repeatedly as in the last paragraph, and we easily get 9.17 (2) at η. This completes the proof of Lemma 9.17. □

We can now easily complete the proof of 9.14. Suppose first that θ is a limit ordinal. For $0 \leq i < \omega$, let α_{i+1} be defined by

$$n^*(\alpha_{i+1}) = \inf\{n^*(\beta) \mid \alpha_i < \beta < \theta\}.$$

(Recall that $\alpha_0 = lh(\mathcal{B}_0) - 1$.) Clearly, $\alpha_i < \alpha_{i+1}$ and α_{i+1} is a break point at θ, for all i. We may suppose that n^* was chosen so that $n^*(\alpha_0) = 0$, which means that α_0 is a break point at θ. But then 9.17 (1) gives us a sequence $\langle \mathcal{F}_i \mid i \in \omega \rangle$ such that $\mathcal{F}_0 = \mathcal{E}_0$, $\mathcal{F}_i$ is a realization of $\Phi(\mathcal{T} \restriction \alpha_i + 1)$, and $\mathcal{R}^{\mathcal{F}_{i+1}}_{\alpha_{i+1}} \in \mathcal{R}^{\mathcal{F}_i}_{\alpha_i}$, for all $i < \omega$. This is a contradiction.

Next, suppose $\theta = \gamma + 1$. We may suppose n^* is chosen so that $n^*(\gamma) = 0$, which implies that β survives at γ whenever $\beta T \gamma$. But then 9.17 (2) clearly implies that $\mathcal{M}^{\mathcal{T}}_\gamma$ is $\mathcal{E}_0$-realizable, as desired. □

Theorem 2.5 obviously follows from 9.14. (While 9.14 was only proved for normal trees, whereas 2.5 was stated for linear compositions of normal trees, we can nevertheless take care of such "almost normal" trees by applying 9.14 (2) repeatedly to their normal components.) The iterability of the exotic creatures of $\mathbb{C}$ which we used in the proof of 1.4 also follows immediately. This represents all the iterability we used in §1 - §5.

It remains only to prove Theorem 6.9, which states that if $K^c \models$ "There are no Woodin cardinals", then every K^c generated phalanx $\mathcal{B}$ such that $lh\ \mathcal{B} < \Omega$ is $\Omega + 1$-iterable. We shall now sketch the minor modifications of the proof of 9.14 which yield this result.

First, the reflection arguments of §2 show that it is enough to prove the following: let $\pi : \mathcal{M} \to K^c$ be elementary, where $\mathcal{M}$ is countable. Let $\mathcal{B}$ be a hereditarily countable phalanx which is $(\Sigma, \mathcal{M})$-generated, where Σ is the strategy of choosing unique cofinal wellfounded branches. Let $\mathcal{T}$ be a countable, normal, putative iteration tree on $\mathcal{B}$. Then either $\mathcal{T}$ has a cofinal wellfounded branch or $\mathcal{T}$ has a last, wellfounded model. So fix π, $\mathcal{M}$, and $\mathcal{B}$ with these properties; we shall show that there is a realization $\mathcal{E}$ of $\mathcal{B}$ such that $\forall \alpha < lh\ (\mathcal{B})$ ($\mathcal{R}^{\mathcal{E}}_\alpha$ has $\delta^{\mathcal{R}^{\mathcal{E}}_\alpha}$ cutoff points). The desired conclusion concerning $\mathcal{T}$ then follows from 9.14.

Since Ω is measurable, we can find $\xi < \Omega$ such that $\pi : \mathcal{M} \to \mathcal{N}_\xi$ is elementary. Let η be such that $(V_\eta, \in, \Omega)$ is a coarse premouse having $\Omega + \Omega$ cutoff points, and let $\mathcal{E}_0 = ((V_\eta, \in, \Omega), \mathcal{N}_\xi, \pi)$. Thus $\mathcal{E}_0$ is a realization of $\mathcal{M}$.

Now fix $\alpha < lh(\mathcal{B})$, and let $\mathcal{S}$ be a countable iteration tree on $\mathcal{M}$ such that $\mathcal{M}_\alpha^{\mathcal{B}}$ is an initial segment of $\mathcal{M}_\gamma^{\mathcal{S}}$, the last model of $\mathcal{S}$, and such that $\mathcal{S}$ has no maximal wellfounded branches. Such a tree $\mathcal{S}$ exists because $\mathcal{B}$ is $(\Sigma, \mathcal{M})$ generated. We have by definition 6.7: if $\alpha + 1 < lh(\mathcal{B})$, then $\forall\gamma(\nu(E_\gamma^{\mathcal{S}}) \geq \lambda(\alpha, \mathcal{B}))$, and if $\alpha + 1 = lh(\mathcal{B})$, then

$$\forall\gamma\forall\beta < \alpha(\nu(E_\gamma^{\mathcal{S}}) \geq \lambda(\beta, \mathcal{B}))\,.$$

We now apply a slight variant of 9.17 to find the desired realization $(\mathcal{R}_\alpha^{\mathcal{E}}, Q_\alpha^{\mathcal{E}}, \pi_\alpha^{\mathcal{E}})$ of $\mathcal{M}_\alpha^{\mathcal{B}}$. Let $n^* : \gamma + 1 \to \omega$ be 1-1 and $n^*(0) = 0$. Let $n : \gamma + 1 \to \omega$ be defined from n^* as in the proof of 9.14, and interpret "survives" and "break point" relative to n as in 9.14. Thus 0 is a break point at γ. For $\beta < \gamma$, and $(\mathcal{R}, Q, \sigma)$ a $\deg^{\mathcal{S}}(\beta)$ realization of $\mathcal{M}_\beta^{\mathcal{S}}$, let $U(\beta, \mathcal{R}, Q, \sigma)$ be defined as in the proof of 9.14: it is the tree of attempts to build a maximal branch b of $\mathcal{S}$ and realize $\mathcal{M}_b^{\mathcal{S}}$ appropriately. Since $\mathcal{S}$ has no maximal wellfounded branches, $U(\beta, \mathcal{R}, Q, \sigma)$ is always wellfounded.

For $\mathcal{F}$ a realization of $\Phi(\mathcal{S} \restriction \xi)$, let us say that $\mathcal{F}$ has *more than enough room* just in case $\forall\beta < \xi(\mathcal{R}_\beta^{\mathcal{F}}$ has $\delta^{\mathcal{R}_\beta^{\mathcal{F}}} + \omega \cdot \text{rank}(U(\beta, \mathcal{R}^{\mathcal{F}}, Q_\beta^{\mathcal{F}}, \pi_\beta^{\mathcal{F}})) + c(\beta, \gamma)$ cutoff points), where of course $c(\beta, \gamma)$ is defined as in the proof of 9.14. So $\mathcal{E}_0$ has more than enough room. It is clear that the proof of 9.17 works equally well when "more than enough room" replaces "enough room" in its hypothesis and conclusion. Since 0 is a break point at γ, this version of 9.17 gives us a realization $\mathcal{F}$ of $\Phi(\mathcal{S})$ such that $\mathcal{E}_0 = \mathcal{F} \restriction 1$ and $\mathcal{R}_\gamma^{\mathcal{F}}$ has $\delta^{\mathcal{R}_\gamma^{\mathcal{F}}}$ cutoff points. Since $\mathcal{F}$ is a realization of $\Phi(\mathcal{S})$, $Q_\gamma^{\mathcal{F}}$ agrees with $Q_0^{\mathcal{F}} = \mathcal{N}_\xi$ below $\sigma \circ \pi(\nu(E_0^{\mathcal{S}}))$ and $\pi_\gamma^{\mathcal{F}} \restriction \nu(E_0^{\mathcal{S}}) = \sigma \circ \pi \restriction \nu(E_0^{\mathcal{S}}) = \sigma \circ \pi \restriction \nu(E_0^{\mathcal{S}})$, where σ is the appropriate complete resurrection embedding (bringing $\pi(E_0^{\mathcal{S}})$ back to life). Now whenever $\beta \leq \alpha$ and $\lambda(\beta, \mathcal{B})$ is defined (i.e. $\beta + 1 < lh(\mathcal{B})$), $\lambda(\beta, \mathcal{B})$ is a cardinal of $\mathcal{M}$ and $\nu(E_0^{\mathcal{S}}) \geq \lambda(\beta, \mathcal{B})$. Thus $\pi(\lambda(\beta, \mathcal{B}))$ is a cardinal of K^c and $\sigma \restriction \pi(\lambda(\beta, \mathcal{B})) + 1$ is the identity. This implies that $\pi_\gamma^{\mathcal{F}} \restriction (\lambda(\beta, \mathcal{B}) + 1) = \pi \restriction (\lambda(\beta, \mathcal{B}) + 1)$, whenever $\beta \leq \alpha$ and $\lambda(\beta, \mathcal{B})$ exists. Let us set $\mathcal{R}_\alpha^{\mathcal{E}} = \mathcal{R}_\gamma^{\mathcal{F}}$, $\pi_\alpha^{\mathcal{E}} = \pi_\gamma^{\mathcal{F}} \restriction \mathcal{M}_\alpha^{\mathcal{B}}$, and $Q_\alpha^{\mathcal{E}} = Q_\gamma^{\mathcal{F}}$ if $\mathcal{M}_\alpha^{\mathcal{B}} = \mathcal{M}_\gamma^{\mathcal{S}}$, and $Q_\alpha^{\mathcal{E}} = \pi_\gamma^{\mathcal{F}}(\mathcal{M}_\alpha^{\mathcal{B}})$ otherwise. Doing this for all $\alpha < lh(\mathcal{B})$, we obtain a realization $\mathcal{E}$ of $\mathcal{B}$ which has enough room; the agreement properties of $\mathcal{E}$ follow from the fact that if $\alpha + 1 < lh\ \mathcal{B}$, then $\pi_\alpha^{\mathcal{E}} \restriction (\lambda(\alpha, \mathcal{B}) + 1) = \pi \restriction (\lambda(\alpha, \mathcal{B}) + 1)$ and $Q_\alpha^{\mathcal{E}}$ agrees with $\mathcal{N}_\xi$ below $\pi(\lambda(\alpha, \mathcal{B}) + 1)$, and from the corresponding facts when $\alpha + 1 = lh(\mathcal{B})$.

From 9.14 we get that the tree $\mathcal{T}$ on $\mathcal{B}$ is well-behaved, and this completes the proof of 6.9.

References

[D] A. J. Dodd, *The core model*, London Math. Soc. Lecture notes, vol. 61, 1982.

[DJ1] A. J. Dodd and R. B. Jensen, *The core model*, Ann. Math. Logic 20 (1981), 43-75.

[DJ2] A. J. Dodd and R. B. Jensen, *The covering lemma for K*, Ann. Math. Logic 22 (1982), 1-30.

[DJ3] A. J. Dodd and R. B. Jensen, *The covering lemma for $L[U]$*, Ann. Math. Logic 22 (1982), 127-155.

[DJKM] A. J. Dodd, R. B. Jensen, P. Koepke, and W. J. Mitchell, *The core model for nonoverlapping extender sequences*, to appear.

[FMS] M. Foreman, M. Magidor, and S. Shelah, *Martin's maximum, saturated ideals, and non-regular ultrafilters*, Ann. of Math. (2) 127 (1988), 1-47.

[H] K. Hauser, *The consistency strength of projective absoluteness*, habilitationsschrift, Ruprecht-Karls-Universitat, Heidelberg, 1993.

[Hj] Greg Hjorth, *Π^1_2 Wadge degrees*, Ann. Pure and Applied Logic 77 (1996), no. 1, 53-74.

[KS] A. S. Kechris and R. M. Solovay, On the relative consistency strength of determinacy hypotheses, TAMS, 290 (1), 179-211.

[Ku1] K. Kunen, *Some applications of iterated ultrapowers in set theory*, Ann. Math. Logic 1 (1970), 179-227.

[Ku2] K. Kunen, *Saturated ideals*, J. Symbolic Logic 43 (1978), 65-77.

[IT] D. A. Martin and J. R. Steel, *Iteration trees*, JAMS 7 (1994), 1-73.

[M1] W. J. Mitchell, *The core model for sequences of measures* I, Math. Proc. Cambridge Philos. Soc. 95 (1984), 228-260.

[M2] W. J. Mitchell, *Σ^1_3 absoluteness for sequences of measures*, in Set Theory of the Continuum, H. Judah, W. Just, H. Woodin eds., MSRI publications, no. 26, Springer-Verlag 1992.

[M?] W. J. Mitchell, *The core model for sequences of measures* II, unpublished.

[FSIT] W. J. Mitchell and J. R. Steel, *Fine structure and iteration trees*, Springer Lecture Notes in Logic 3 (1994).

[WCP] W. J. Mitchell, E. Schimmerling and J. R. Steel, *The weak covering lemma up to a Woodin cardinal*, to appear in Ann. Pure and Applied Logic.

[Sch] E. Schimmerling, *Combinatorial principles in the core model for one Woodin cardinal*, Ann. of Pure and Applied Logic, 74 (1995) 153-201.

[TM] E. Schimmerling and J. R. Steel, *Fine structure for tame inner models*, J. Symbolic Logic, vol. 61 (1996), 621-639.

[CMWC] J. R. Steel, *Core models with more Woodin cardinals*, to appear.

[PW] J. R. Steel, *Projectively wellordered inner models*, Ann. of Pure and Applied Logic, vol. 74 (1995), 77-104.

[SVW] J. R. Steel and R. Van Wesep, *Two consequence of determinacy consistent with choice*, Trans. of AMS 272 (1982), 67-85.

[SW] J. R. Steel and P. D. Welch, Σ^1_3 *absoluteness and the second uniform indiscernible*, to appear.

[To] S. Todorčevic, *A note on the proper forcing axiom*, Contemporary Mathematics 95 (1984), 209-218.

[W] W. H. Woodin, *Some consistency results in ZFC using AD*, in Cabal Seminar 79-81, Springer Lecture Notes in Mathematics, vol. 1019 (1983), 172-198.

Index of Definitions

Definitions not numbered in the text are indexed here by the number of the theorem, lemma, or definition immediately preceding; thus "1.4 ff." indicates an unnumbered definition occuring in the body of the text after Theorem 1.4.

Lightning Source UK Ltd.
Milton Keynes UK
UKOW02n0855130317
296147UK00019B/175/P

9 781107 167964